CHEMICAL REACTIONS IN LIQUID AND SOLID PHASE: KINETICS AND THERMODYNAMICS

CHEMICAL REACTIONS IN LIQUID AND SOLID PHASE: KINETICS AND THERMODYNAMICS

G. E. ZAIKOV AND A. JIMÉNEZ
EDITORS

Nova Science Publishers, Inc.
New York

Senior Editors: Susan Boriotti and Donna Dennis
Coordinating Editor: Tatiana Shohov
Office Manager: Annette Hellinger
Graphics: Wanda Serrano
Editorial Production: Maya Colmbus, Vladimir Klestov,
Matthew Kozlowski and Tom Moceri
Circulation:, Raymond Davis, Melissa Diaz, Ave Maria Gonzalez, Magdalena Nuñez,
Marlene Nuñez and Jeannie Pappas, Vera Popovich
Communications and Acquisitions: Serge P. Shohov
Marketing: Cathy DeGregory

Library of Congress Cataloging-in-Publication Data
Available upon request

ISBN 1-59033-854-5

400 Oser Ave, Suite 1600
Hauppauge, New York 11788-3619
Tele. 631-231-7269Fax 631-231-8175
e-mail: Novascience@earthlink.net
Web Site: http://www.novapublishers.com

Printed in the United States of America

CONTENTS

PREFACE

G. E. Zaikov and A. Jiménez

One of the principal objects of theoretical
research in any field of knowledge
is to find the point of view from which the
subject appears in its greatest simplicity
J.Williard Gibbs
(One of the most important scientists explaining
thermodynamic theory)

It is well known from school textbooks that a compound can exist in gas, liquid and solid phases as well as plasma. All these conditions are radically differentiated. That is why different chemical reactions go in different ways. The primary attention is directed particularly at condensed phase (liquid and solid states) in this volume. Various articles examine the influence of liquid and solid states during radical, ionic and molecular reactions. Cage effect, diffusion hindrance, donor-acceptor interaction, electrostatic interaction, dispersion forces and many other factors greatly influence the rates, mechanism and direction of chemical reactions [1, 2].

In spite of the enormous number of papers dealing with the influence of the medium on the rate of chemical reactions, no strict quantitative theory capable of "universal" application has been put forward up to now. It is thus not yet possible to describe the relationship between the reaction rate constants and the equilibrium constants with the nature of the medium in which the reactions are taking place by means of a single equation.

The absence of a general theory on the influence of the medium on the kinetics of a chemical reaction can be explained by the fact that a change of solvent (or transfer from liquid to solid phase) may not only influence the rate of the process but frequently also results in a complication of the reaction mechanism. The calculation of the individual contributions made by each factor is thus, in most cases, rather complicated and requires a deep and comprehensive study of the properties of the medium and of the reacting particles. This is because a quantitative evaluation of all types of interactions between the reacting particles with medium can only be arrived at on basis of full knowledge of these properties.

Although it is possible, in principle, to obtain detailed information about a process and to solve the problem of the change in entropy and in the activation energy occurring during the interaction of reacting particles with the medium, it is, at present, impossible, in practice, to arrive at an appropriate formulation of the problem, quite apart from solving it, on account of the absence of a number of quantitative process characteristics (frequently qualitative ones as well).

We would like to note that the interactive forces between molecules in the gaseous state are extremely small, hence the molecular constitution in the vapour phase can be described fairly accurately by static relationships. A description of the liquid state is considerably more difficult because it has neither a fully statistical distribution (which is a characteristic of the gaseous state) nor a fully regular structure (which is a characteristic of the solid state, especially crystalline one). Therefore, to describe any kinetic processes in the liquid phase, it is often advisable to use less fundamental relationships than in the case of vapour phase reaction.

It is even more complicated with solid state reactions where molecules can be organized in very versatile way (from the complete chaous to molecule's full organization).

The given volume presents examples on kinetics and mechanism of chemical reactions in liquid and solid states.

REFERENCES

[1] N.M.Emanuel, G.E.Zaikov, Z.K.Maizus, *Oxidation of Organic Compounds. Effect of Media*. Pergamon Press, Oxford, 1984, 526 p

[2] N.M.Emanuel, G.E.Zaikov, V.A.Kritsman, *Chemical Kinetics and Chain Reactions*. Historical Aspects, Nova Science Publishers, NY, 1995, 492 p

Prof. Gennady E. Zaikov
Institute of Biochemical Physics Russian Academy of Sciences
Moscow, Russia
Prof. Alfonso Jimenez
University of Alicante
Alicante, Spain

In: Chemical Reactions in Liguid and Solid Phase
Editors: G. E. Zaikov and A. Jiménez, pp. 1-10

ISBN 1-59033-854-5

Chapter 1

SWELLING BEHAVIOR OF POLYMERIC NETWORKS BASED ON N-ISOPROPYLACRYLAMIDE AND N, N-DIMETHYLACRYLAMIDE HYDROGELS

D. García[a*], J. L. Escobar[a], N. Bada[a] and I. Katime[b]

[a] Departamento de Macromoléculas, Centro de Biomateriales, Universidad de la Habana, 10400, Havana, Cuba.

[b] Grupo Nuevos Materiales, Departamento de Química Física, Facultad de Ciencias, Campus de Leioa, Universidad del País Vasco, 644, Bilbao, Spain.

ABSTRACT

Cross-linked films of N-isopropylacrylamide (NIPAAm) and N, N-dimethylacrylamide (DMAAm) hydrogels with different compositions were prepared by free radical copolymerization initiated by potassium persulfate ($K_2S_2O_8$) at 60°C. The swelling behavior of film prepared with 0.88, 0.62, 0.44 and 0.26 mol % of NIPAAm was followed gravimetrically and the diffusion coefficients were determined according to the Fickian law at several temperatures in the interval 20-40^{0}C. It was observed that the water uptake in the copolymers decreases with the NIPAAm content decrease. Cefazolin release has been examined as a function of time. The release experiments were carried out at 37°C. The fraction of available drug release was linear in $(\text{time})^{1/2}$ during the initial state. These studies allow to determinate a diffusion coefficient for cefazolin release from the copolymeric hydrogels.

INTRODUCTION

Hydrogels are insoluble, crosslinked polymer network structures composed of hydrophilic homo- or hetero-co-polymers, which have the ability to absorb significant amounts of water. From a biological viewpoint, this is an essential property to achieve an

* Corresponding author, email: jlescobar@biomat.uh.cu

immunotolerant surface and matrix (i.e., with respect to protein adsorption or cell adhesion). Due to their significant water content, hydrogels also possess a degree of flexibility very similar to natural tissue, which minimizes potential irritation to surrounding membranes and tissues [1]. Hydrogels swollen in an aqueous solution sometimes exhibit a discontinuous volume change in response to changes in environmental conditions such as temperature, solvent composition, pH, etc. This phase change has been denoted the volume phase transition or gel-gel transition of gels [2].

Poly(N-isopropylacrylamide) (NIPAAm) is in the class of polymers which exhibit lower critical solubility temperature (LCST) behavior [3-5]. Linear and crosslinked thermo-sensitive hydrogels form a single phase with water and swell at low temperatures, and collapse to form a separate phase above the critical temperature. Because the critical temperature of collapse for NIPAAm is approximately 32°C, this polymer has potential as a sensor in humans, as well as in size-selective separation processes involving materials, which can be denatured at high temperatures [6]. Poly(N isopropylacrylamide) has the most sharp swelling transition of the class of thermo-sensitive alhyl acrylamide polymers [7], making it extremely sensitive to small changes in temperature.

In this paper, poly (N-isopropylacrylamide-co-N,N-dimethylacrylamide) copolymers were synthesized by radical copolymerization in aqueous solution at 60°C, using potassium persulfate as initiator. The temperature influence on the swelling process of copolymeric hydrogels with different compositions was determined. The diffusion coefficients for each copolymer were analyzed on the basis of Fick's diffusion theories. Release kinetic of cefazolin from glassy as well as swollen N-isopropylacrylamide-co-N,N-dimethylacrylamide crosslinked hydrogels in aqueous solutions have been studied.

EXPERIMENTAL

Materials

N-isopropylacrylamide (NIPAAm) (Fluka; HPLC grade), N,N-dimethylacrylamide (DMAAm) (Fluka; HPLC grade), potassium persulfate ($K_2S_2O_8$) (MERK; HPLC grade) and Cefazolin (Pharmaceutic Laboratories Reynaldo Gutiérrez) were used as received. N,N-methylenebisacrylamide (NNMBAA) (Fluka; HPLC grade) was recrystallized from a methanol solution. NNMBAA was employed as crosslinking agent. Doubly distilled and deionized water with a conductivity 0.1 μs was employed in the swelling studies.

Polymerization Procedure

The hydrogels were synthesized by radical polymerization. The feed mixtures are given in Table 1. The initiator was $K_2S_2O_8$ (1.0 wt %) and the crosslinking agent was NNMBAA (see Figure 1) (0.25 wt % with respect to the total weight of the monomers). Doubly distilled and deionized water was used as solvent. The solution was outgassed with gaseous nitrogen in glass tubes. Copolymerization was carried out at 60°C for 8 h.

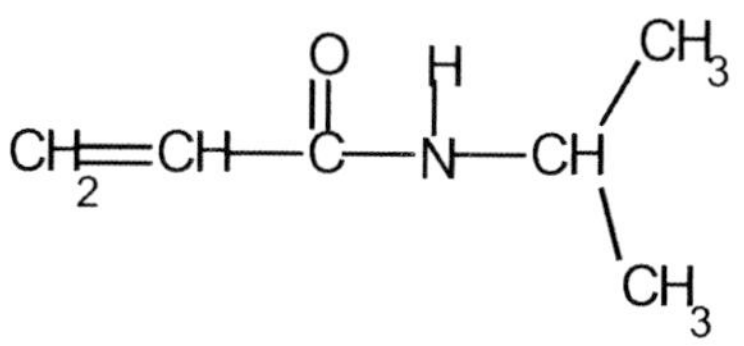

N-isopropylacrylamide

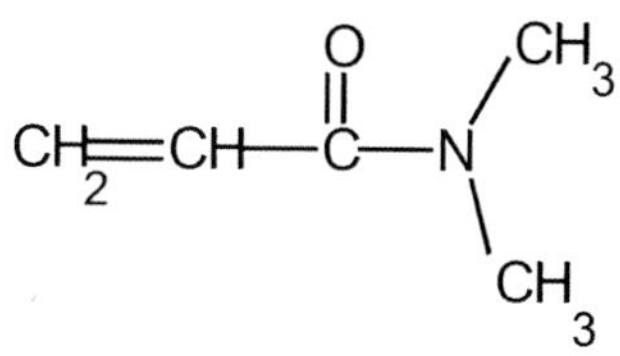

N,N′-dimethylacrylamide

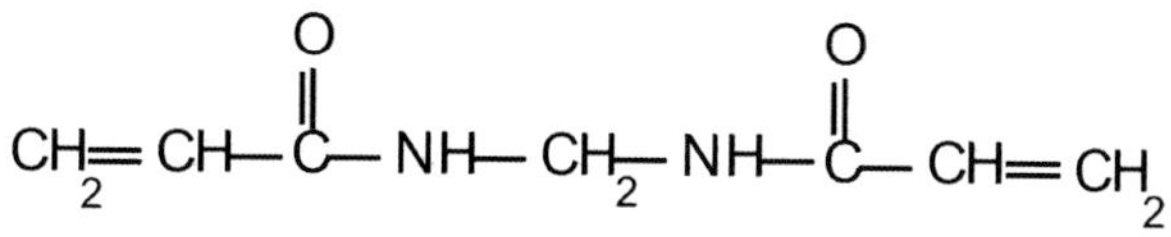

N,N′-methylenebisacrylamide

Figure 1. Chemical structures of the monomers unit and the crosslinking agent.

Table 1. Composition Data of Reaction Mixture.

Samples	NIPAAm content (mol %)	DMAAm content (mol %)
M_1	0.88	0.12
M_2	0.62	0.38
M_3	0.44	0.56
M_4	0.26	0.72

The resultant crosslinked polymer rods (see Figure 2a) were immersed in water for two weeks to remove any possible residual monomers. Discs were obtained from the hydrogels rods. They were dried at ambient temperature for several days and then in an oven at 40°C until constant weight. The thickness of the xerogel (dry disc) was 1.00 ± 0.01 mm (see Figure 2b).

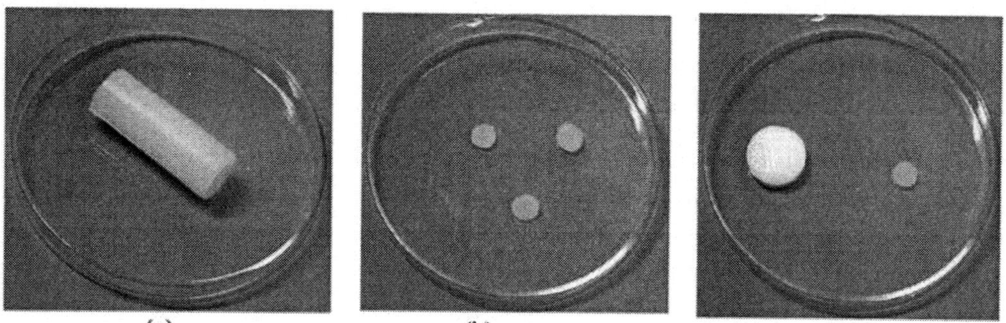

(a) (b) (c)

Figure 2. Photograph of exposed crosslinked copolymer rod (a), xerogels (b) and the difference of the dimensions between swollen disc (left) and xerogel (right).

Swelling Experiment

Dynamic swelling experiments were performed by placing the glassy discs in a deionised and doubly distilled water bath at several temperatures (20, 30, 32, 34, 37 and 40±0.1°C). Figure 2(c) shows the difference of the dimensions between swollen disc (left) and xerogel (right). The water uptake (swelling), W, was calculated by measuring their weight gain of the sample at different times. It was reported as:

$$W = \frac{(\text{weight of swollen disc - weight of dry disc})}{\text{weight of dry disc}} \tag{1}$$

Loading of Hydrogels

Polymer discs were loaded with cefazolin (see Figure 3) by soaking them is an aqueous solution of the drug (2.26 x 10^{-3} g/L). The swollen drug-loaded gels were then either dried at ambient temperature for several day and then at 40°C until constant weight. The dry drug-loaded gels were used in release experiments.

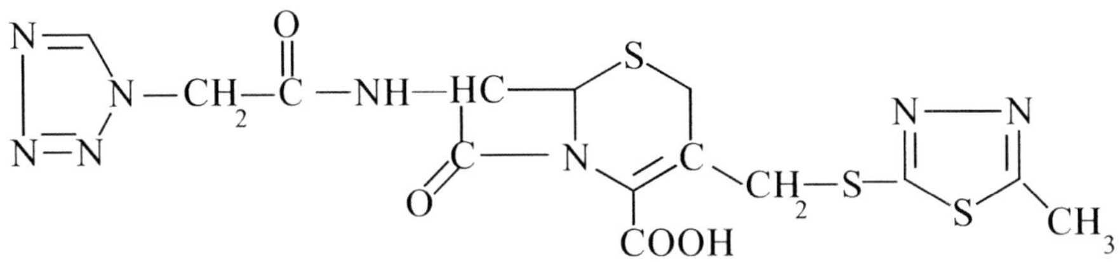

Figure 3. Chemical structure of Cefazolin

Drug Release Experiments

Release experiments were performed by placing the polymeric disc containing the drug on a holder into to distilled water at 37±0.1°C.

The volume of water in the pyrex tubes glass was 10 mL. To follow the cefazolin release, the water present in the pyrex tubes glass was extracted after having introduced the drug-polymer conjugated a various times. Later 10 mL of distilled water in the same pyrex tubes glass to follow the release study was added again. The concentration of cefazolin was measured by U.V spectrophotometer (secoman, S.1000 model) and the wavelength detect was 261 nm.

The amount of drug release at any selected time was calculated from the cefazolin calibration curve.

RESULTS AND DISCUSSION

Swelling Experiments

Swelling studies were performed gravimetrically following equation (1). The values of equilibrium degree of swelling, W_∞, at different temperatures are reported in Table 2.

Table 2. Values of the Equilibrium Degree of Swelling, W_∞, for the NIPAAm/DMAAm Hydrogels at Several Temperatures.

		W_∞					
Samples	NIPAAm content (mol %)	T=20°C	T=30°C	T=32°C	T=34°C	T=37°C	T=40°C
M_1	0.88	8.40	4.50	1.7	0.9	0.7	0.7
M_2	0.62	11.6	8.40	7.8	6.9	5.3	5.2
M_3	0.44	13.8	13.3	12.7	12.3	10.3	10.4
M_4	0.26	14.9	15.4	15.9	13.6	18.3	18.3

Table 2 shows the equilibrium swelling degree values at different temperatures as a function of the swelling time for four NIPAAm/DMAAm copolymers. As can be observed, there is a strong dependence of the swelling on compositions, which is reflected in the equilibrium swelling degree. The swelling of crosslinked copolymer is the result of a compromise between the osmotic swelling pressure and the elastic force of the macromolecular chains of the network opposing it. Therefore, as the composition of NIPAAm/DMAAm copolymers becomes richer in NIPAAm, the equilibrium swelling degree decreases.

On the other hand, the effect of temperature variations on equilibrium swelling degree was also analyzed. As you can see in Figure 5, the equilibrium swelling degree exhibit a lower critical solubility temperature (LCST) about 33°C in water (the volume decreases at temperatures higher than 33°C) (see Figure 4), for copolymer compositions rich in NIPAAm.

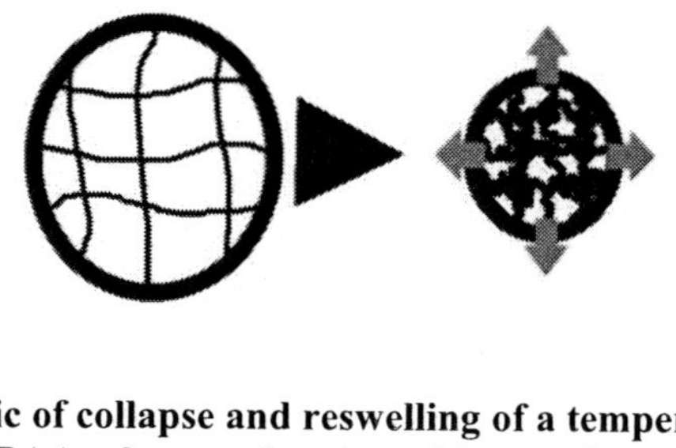

Figure 4. Schematic of collapse and reswelling of a temperature-sensitive of NIPAAm homopolymer and its copolymers.

These behaviors are in a very good agreement with those recently reported for NIPAAm homopolymer and its copolymers whit methacrylic acid [8] and acrylic acid [9], which presents a LCST between 32 and 35°C.

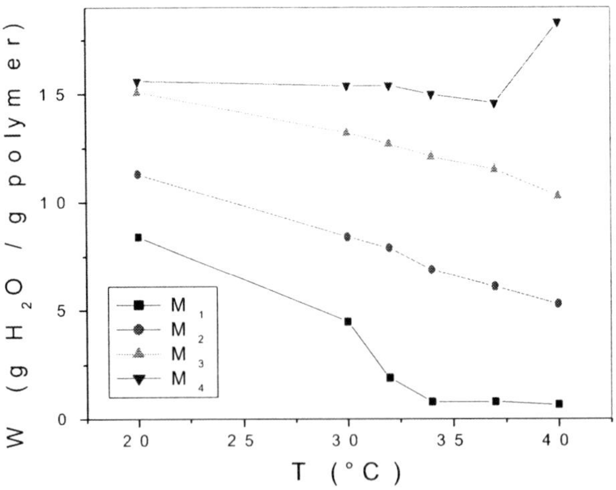

Figure 5. Swelling isotherms for NIPAAm/DMAAm copolymers at 20, 30, 32, 34, 37 and 40°C.

The penetration of solvent into a stiff polymer sample is usually described as the result of two different processes: the diffusion of the solvent into the swollen matrix, and the advancement of the swollen-unswollen boundary as a consequence of the stress induced in the polymer [10]. When the rate-determining step is the first mechanism, there is a linear dependence between the equilibrium degree of swelling and $t^{1/2}$, and the system is said to exhibit a Fickian behavior. In contrast, if the advancement of the swollen-unswollen boundary is slower than the diffusion of the solvent in the swollen polymer, zero-order kinetics are observed and the equilibrium degree of swelling increases linearly with the swelling time.

The solution of differential form of Fick's law for thin sheets neglecting diffusion through the edges, can be expressed as a function of swelling degree M_t / M_∞ (or W_t / W_∞) and $t^{1/2}$ according to the following relationship [11]:

$$\frac{M_t}{M_\infty} = 4\sqrt{\frac{D}{\pi l^2}} \tag{2}$$

where M_t y M_∞ represents the swelling degree at time t and equilibrium, respectively, D is the diffusion coefficient and l is the average thickness of the film.

From the straight lines of the M_t / M_∞ versus $t^{1/2}$ plots, and taking into consideration the value of W_∞ reported in Table 2 and the thickness of the film, the diffusion coefficient of copolymers were determined. The values obtained are reported in Table 3.

It can be appreciated that Fickian behavior is obeyed up to approximately 180 min of swelling, although for longer times the behavior deviates from the linearity. The pattern was obtained for the experiments performed at 20, 30, 32, 34, 37 and 40°C.

Table 3. Values of Diffusion Constants, *D*, for the NIPAAm/DMAAm Hydrogels at Several Temperatures.

Samples	NIPAAm content mol(%)	D x 10^7 cm^2/seg 20°C	30°C	32°C	34°C	37°C	40°C
M_1	0.88	7.30±0.01	5.01±0.01	4.89±0.02	4.56±0.01	4.03±0.01	3.90±0.01
M_2	0.62	7.80±0.01	5.20±0.01	5.02±0.02	4.78±0.02	4.23±0.01	4.20±0.01
M_3	0.44	8.60±0.01	5.70±0.01	5.40±0.01	5.26±0.01	5.06±0.01	4.95±0.01
M_4	0.26	9.10±0.02	8.50±0.01	6.32±0.01	5.86±0.01	5.16±0.02	5.46±0.02

There is noticeable difference in the diffusion coefficients of NIPAAm/DMAAm copolymer, but the copolymer with 0.88 mol% of NIPAAm exhibit a small values of *D*.

For long swelling times the experimental points seem to deviate from the Fickian behavior. For extensive swelling of polymers, Schott [12] proposed an expression of the form:

$$\frac{t}{W} = A + Bt \tag{3}$$

where *W* if the water uptake at time t, $B=1/W\infty$, the inverse of the maximum swelling and $A=1/(dW/dt)_0$, the reciprocal of the initial swelling rate. In fact, Schott demonstrated that equations (3) implies a second-order swelling kinetics.

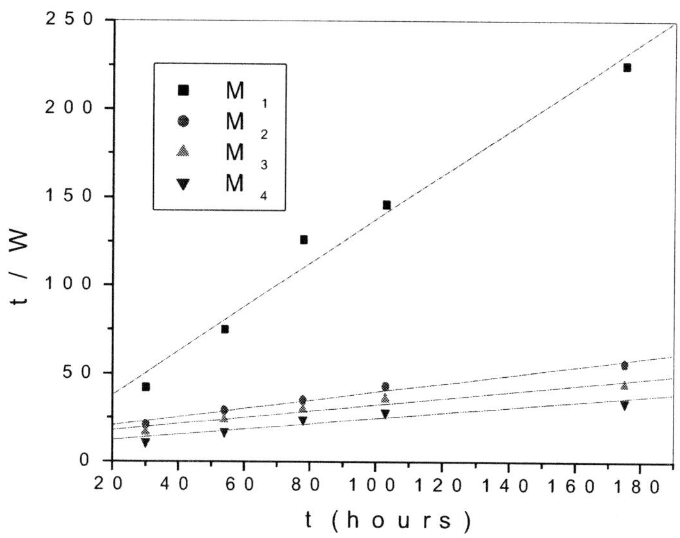

Figure 6. Reciprocal of the average swelling rate t/W, as a function of the time of treatment. Straight lines correspond to the least-squares fit of experimental points to equation (3).

The diagrams obtained by the application of equation (3) give straight lines with excellent correlation coefficients, as is illustrated in Figure 6 for the swelling data obtained at 37°C. The same pattern was obtained for the data at 20, 30, 32, 34 and 40°C.

Drug Release

The kinetic of cefazolin release from the glassy poly(NIPAAm-co-DMAAm) hydrogels with different compositions are presented in Figure 7.

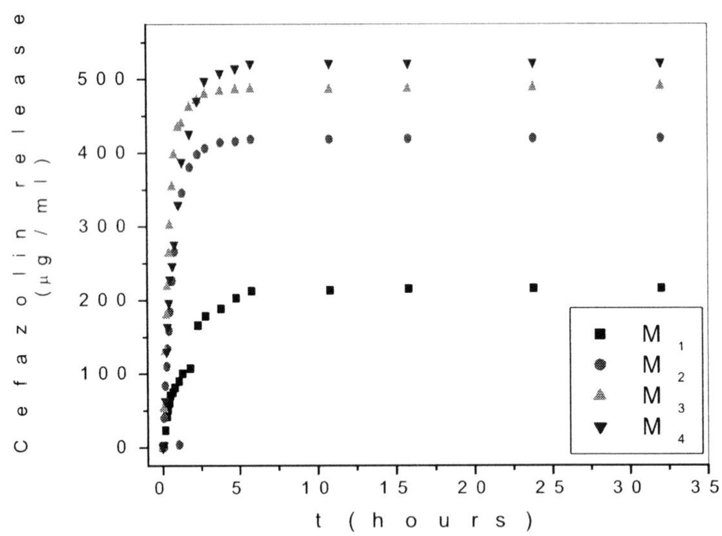

Figure 7. Cefazolin release curve from glassy poly(NIPAAm-co-DMAAm) hydrogels with different compositions.

As can be seen in Figure 7, the cefazolin release depends on the NIPAAm content. A higher content of NIPAAm results in a slower release rate. The cefazolin release percent shown in Table 4.

Table 4. Cefazolin Release Percent from Poly(NIPAAm-co-DMAAm) Hydrogels at 37 °C.

Samples	Cefazolin release percent
M_1	22.0±0.1
M_2	42.1±0.1
M_3	49.0±0.1
M_4	52.8±0.1

The solute release data may be analyzed with equation (4):

$$\frac{Mt}{M\infty} = kt^n \tag{4}$$

where, in this case, $\frac{Mt}{M\infty}$ is the fractional solute release, t is the release time, k is a kinetic constant characteristic of the drug/polymer system, and n is an exponent which characterizes the mechanism of release of the drug.

This equation describes the release kinetics of drugs, which diffuse by Fickian mechanism. For this case n =0.5 and k *is* given by

$$k = 4\left(\frac{D_i}{\pi l^2}\right)^{1/2} \tag{5}$$

where D is the diffusion coefficient for drug delivery and l is the thickness of the disc containing the drug.

The diffusion coefficient of cefazolin in equilibrium swollen was measured by drug release experiments from the crosslinked swollen gels in a solution of cefazolin. The diffusion coefficient was calculated from the slope of the release curve according to follow equation:

$$D = \frac{\pi l^2}{16}\left[\frac{d(Mt/M\infty)}{dt^{1/2}}\right]^2 \tag{6}$$

Values for the k, n and diffusion coefficient (D) of the cefazolin gels are listed in Table 5.

Table 5. The Values of *n*, *k* and Diffusion Coefficient (*D*) of the Cefazolin Release Kinetic.

Samples	NIPA content mol(%)	n	$k \times 10^2$ (horas^{-n})	$D \times 10^7$(cm^2/ seg)
M_1	0.88	0.49	63.79± 0.02	43.35± 0.02
M_2	0.62	0.48	65.74 ± 0.02	49.20 ± 0.02
M_3	0.44	0.51	71.40 ± 0.01	52.90 ± 0.05
M_4	0.26	0.50	73.26 ± 0.01	56.82 ± 0.04

The release index obtained for release kinetics from the swollen hydrogels (see Table 5) suggest that this is a diffusion controlled process.

ACKNOWLEDGEMENTS

Financial support for this work was provided by CYTED project.

REFERENCES

[1] Byrne, M.E., Park, K. and Peppas, N.A., *Advanced Drug Delivery Reviews, 54* 149-161 (2002).

[2] Peppas, N.A., *Hydrogel in Medicine and Pharmacy*. CRC Press, Boca Ratón, FL, (1987).

[3] Brannon-Peppas, L. And Peppas, N.A., *Journal Controlled Release, 16* 319-330 (1991).

[4] Heskins, M. And Guillet, J.E., *J. Macromol. Sci-Chem.*, 1441-1455 (1968).

[5] Chiklis, C.K. and Grasshoff, J.M., *J. Polym. Sci.*, 1617-1626 (1970).

[6] Frietas, R.F.S. and Cussler, E.L., *Chem. Eng. Sci., 42* 97-103 (1987).

[7] Gutowska, A., Bae, Y.H., Feijen, J. And Kim, S.W., *Journal Controlled Release, 22* 95-104 (1992).

[8] Brazel, C., Peppas, N.A., *Journal Controlled Release, 39* 57-64 (1996).

[9] Jeong, B., Wam Kim, S., Han Bae, Y., *Advanced Drug Delivery Reviews, 54* 37-51 (2002).

[10] Montheard, J.P., Chatzopoulos, M. and Chappard, D., *J. Macromol. Sci. Reviews, 32* 1 (1992).

[11] Crank, J., *The Mathematics of Diffusion*. Clarendon, Oxford, *11* 203 (1984).

[12] Schott, H., *J. Macromol. Science-Physics, 1* B31 (1992).

In: Chemical Reactions in Liguid and Solid Phase
Editors: G. E. Zaikov and A. Jiménez, pp. 11-59
ISBN 1-59033-854-5

Chapter 2

HYDROGELS IN ENDOVASCULAR EMBOLIZATION

D. Horak, K. Z. Gumargalieva[1] and G. E. Zaikov[2]
Institute of Macromolecular Chemistry, 26 Heirovskogo st., Prague-6, Czeck Republic
[1]N.N. Semenov Institute of Chemical Physics,
4 Kosygin st., Moscow 119991, Russia
[2]N.M. Emanuel Institute of Biochemical Physics,
4 Kosygin st., Moscow 119991, Russia

1. SPHERICAL PARTICLES OF POLY(2-HYDROXYETHYL METHACRYLATE) AND THEIR MEDICO-BIOLOGICAL PROPERTIES

1.1 Introduction

Endovascular occlusion of blood vessels is an innovative procedure in surgical practice; it is used to stop bleeding, reduce blood supply to various organs and parts of the body, and the like. However, its application on a major scale is impeded by the restricted choice of suitable materials; moreover, there is no special material specifically suited to endovascular occlusion.

Up to now, particles of various materials have been employed in vascular occlusion, such as metals, including ferromagnets, natural and synthetic polymers, e.g. collagen sponge, polyacrylates, polyurethane, polytetrafluoroethylene, cyanoacrylate adhesives (which remain permanently in the organism) or even an autologous precipitate of the organism's own blood which, after some time, is resorbed or degraded to soluble components. However, these materials display a number of shortcomings, such as difficulty in processing, electrolytical corrosion and inflexibility of metals, the possibility of malignant growth and also hydrophobicity which in the case of synthetic polymers make them behave in the blood vessel as a completely foreign body, in a similar way to rare metals for example. It should be also pointed out that sufficient purity can never can be guaranteed. In view of the hydrophilicity of living matter, it is evident that hydrophilic synthetic polymers based on 2-hydroxyethyl methacrylate can remove these shortcomings. Due to their hydrophilicity and high swelling ability, these polymers are soft and permeable to water, which facilitates quick removal of all low-molecular weight, water soluble compounds from the polymer [1-3], and consequently its purification. Long-term experience gained in the field of reconstruction and plastic surgery using homogeneous and microporous hydrogels in particular, has revealed their advantages: nontoxicity, nonirritability, good "healing-in", antigenicity, excellent compatibility with tissues of the living organism, and also, the facts that they do not cause febrile reactions and are not subject to fast destruction [4]. Moreover, poly(2-hydroxyethyl methacrylate) is widely applied in the production of soft contact lenses [1].

Although it seemed at first that hydrogels used as part of vascular prostheses were thromboresistant [5], it appeared later that in some cases they behaved to some extent as thrombogenic material [6] which could advantageously be used in embolization.

The first approach to the synthesis of particles for endovascular occlusion based on poly(2-hydroxyethyl methacrylate) consisted in bulk polymerization in a cylindrical mould; a spaghetti-like product [7] was obtained, which for application in occlusion was cut into short cylinders which were then introduced by operation. They must be spherical to make a simple transcatheteral introduction of particles possible for endovascular occlusion. This includes treatment of haemoptysis and angiodysplasia, occlusion of bronchial, renal, rectal, hepatic and other incoming arteries in vascular surgery, as a primary condition in the preoperative preparation of patients before the excision of vascular tumors and in those cases where, due to the localization of the vascular tumor, its removal is impossible.

In this Chapter we describe a new approach to the synthesis of spherical macroporous poly(2-hydroxyethyl methacrylate) particles and discuss their physicochemical and medicobiological properties.

1.2 Materials and Methods

1.2.1 Materials

Monomers and Auxiliary Compounds

The chemicals were redistilled before use: ethylene dimethacrylate (Ugilor S.A., France) - b.p. 50°C/7 Pa; cyclohexanol (Lachema Brno) - b.p. 68°C/1.6 kPa. Kerosine, decalin, tetralin, butyl acetate, methacrylic acid were also distilled. 2,2'-Azo(bis-isobutyronitrile) (Ferak, Berlin) was recrystallized twice from ethanol. 2-Hydroxyethyl methacrylate (Leciva Modrany), medical purity, and 1-dodecanol (Fluka A.G. Switzerland) were used without purification.

Poly(vinyl pyrrolidone) K-90 (Fluka A.G.), poly(vinyl alcohol) (Polyviol W 25/140, Wacker, FRG) and the other chemicals were reagent grade.

1.2.2 Methods

1-Dodecyl methacrylate (Fluka A.G.) was polymerized in the form of a solution in decalin, with 0.005 wt.%, 2,2'-azo(bis-isobutyronitrile) as initiator, in inert atmosphere in an ampoule at 70°C for 30 h.

Preparation of Hydrogel Particles

Polymer particles were synthesized by the suspension radical copolymerization of 2-hydroxyethyl methacrylate with ethylene dimethacrylate, and/or methacrylic acid in a laboratory reactor (250 ml volume) provided with an anchor-type stirrer. The mixture of monomers, together with an inert solvent and the initiator of the radical polymerization, were dispersed by vigorous stirring in the dispersing medium containing a dissolved suspension stabilizer which prevents agglomeration of particles during stirring. The optimal particles used in the medico-biological tests were obtained by the following procedure: dispersed organic phase consisting of 9.8 ml 2-hydroxyethyl methacrylate, 0.2 ml ethylenedimethacrylate, 7.5 ml 1-dodecanol, 7.5 ml cyclohexanol and 0.1 g 2,2'-azo(bis-isobutyronitrile); 1% water solution of poly(vinyl pyrrolidone) K-90 served as a dispersing medium. The polymerization proceeded at 70°C for 8 h under continuous stirring (150 rpm). After cooling the polymer was separated into the beaker and decanted 10 times each by water and methanol respectively, 3 times each by acetone and ether respectively (each portion contained ca. 300 ml of solvent and was left stationary for at least 4 h). After drying in vacuum the spherical macroporous particles were sieved into separate fractions: 0.4 - 0.6; 0.6 - 0.8; 0.8 - 1.0; 1.0 - 1.2 and 1.2 - 1.5 mm. The particle size distribution curve is obtained from weights of the respective fractions. For the investigation described in this paper, fraction 0.6 - 0.8 mm was used.

Before use, 1 ml of particles was immersed in 250 ml water which was changed daily for 5 days, when, (according to the UV spectrum), the concentration of the low-molecular weight impurities in the solution was lower than that obtained by washing-out 10^{-5} g of impurities from 1 g of dry polymer. Sterilization of samples was carried out in boiling distilled water for 45 min.

Porosity (% of pore volume related to the volume of polymer particles) was calculated from the increase in the diameter of the particles due to the equilibrium swelling of hydrogel beads in water, at room temperature, measured by a GM-1 horizontal microscope equipped with an ocular micrometer.

The specific surface area was measured from the dynamic desorption of nitrogen according to BET isotherm; the sorption of water vapour by the hydrogel was measured using a McBain's balance. The diffusion coefficient was evaluated according to the next Equation [8]:

$$\frac{\mathbf{M_t}}{\mathbf{M_\infty}} = \frac{6}{\mathbf{r}} \times \left(\frac{\mathbf{Dt}}{\pi}\right)^{0.5} \qquad (1)$$

where $\mathbf{M_t}$, $\mathbf{M_\infty}$ are the amounts of sorbed water in time **t** and at the end of the sorption, respectively; **D** is the diffusion coefficient, and **r** the particle radius.

Changes in the morphological picture of blood were quantified under the microscope in 1 ml samples prepared as a mixture of 0.1 ml water extract from hydrogel and 0.9 ml donor blood in the cell.

Implantation

The hydrogel samples were intravascularly and subcutaneously implanted in rabbits. After preparation of the operation zone and local Novocain anesthesia, a linear incision was made along the projection of the vessel leash. When the major femoral arteries were isolated, the catheter was installed into the vascular lumen. Then the emboli were introduced into the vessel through the catheter. In order to check their location, angiography (Urografin 76%, 3 ml) and X-ray images were performed prior to and after the embolization.

If the emboli were subcutaneously introduced into the inguinal zone, X-ray images were taken and sutures were made.

After a certain time (2 weeks, 5 weeks or 21 months), samples of the surrounding tissue were cut out and pieces of the tissue were embedded in celloidin. Histological preparations were stained with haematoxylin and eosin.

1.3 Results and Discussion

1.3.1 Synthesis of Particles for Embolization

Suspension radical polymerization is the optimal method for the preparation of spherical polymer particles [9]. The principle is that the dispersed monomer phase is stirred and polymerized in a dispersing medium immiscible with it. At the same time, the monomer mixture which was stirred into spherical drops must be protected (stabilized), by means of a suspension stabilizer to prevent agglutination or agglomeration. The stabilizer differs in its chemical character, depending upon whether the medium is water or an organic liquid.

If the synthesized beads are to be suitable for the purposes of endovascular occlusion, they have to be soft, possess a sufficient swelling ability and at the same time have satisfactory mechanical properties. For this reason, the suspension polymerization of 2-hydroxyethyl methacrylate with a small amount of the crosslinking agent (2 wt.%) was systematically investigated, and the effect of inert components on the product's properties studied. The presence of inert solvents is a condition for the formation of a heterogeneous porous structure; the fibrous tissue grows into the pores, thus making anchoring of the gel in the vascular lumen possible.

Since 2-hydroxyethyl methacrylate is a hydrophilic monomer readily soluble in water, the inverse phase polymerization was tested first. The monomer dissolved in water as an inert solvent, was polymerized by using ammonium persulphate (initiator) after dispersing it in an organic liquid immiscible with water, i.e. decalin, medical paraffin oil, kerosine, or tetralin. Only with decalin spherical particles, up to 2 mm in size, were obtained. Preparation of a suitable suspension stabilizer for this system, i.e. poly(1-dodecyl methacrylate) having a molecular weight 4×10^6, is however time-consuming, while other stabilizing systems, e.g. based on cetyl alcohol, Span 85 or butyl rubber, were found to be unsuitable.

For this reason, only polymerization in water as the dispersing medium was studied; the effect of composition, both of the dispersed phase and of the dispersing medium, on the properties of the resulting product were examined. A number of stabilization systems were tested, such as xanthan gum with hydroxypropylmethyl cellulose (also in a saturated sodium chloride solution), sodium polymethacrylate, poly(vinyl alcohol), and gelatine. Poly(vinyl pyrrolidone), molecular weight 360 000, was found to be best for this purpose.

With water used as the dispersing medium, the danger arises that the monomer might dissolve in the water. This may be prevented by choosing inert solvents, added to the dispersed phase, with which the monomer has a much higher distribution coefficient than with water, and thus does not pass into it. The mixture of 1-dodecanol and cyclohexanol appeared to be especially suitable. With cyclohexanol alone, a nonporous, hard, glassy polymer was obtained. Other inert systems, such as, e.g. butyl acetate or polystyrene, were equally unsuitable.

Only particles of poly(2-hydroxyethyl methacrylate) crosslinked with 2% ethylene dimethacrylate and prepared in the presence of a mixture of cyclohexanol and 1-dodecanol in an aqueous solution of poly(vinyl pyrrolidone) were used in the experiments.

1.3.2 Physico-Chemical Properties

Particles of poly(2-hydroxyethyl methacrylate) prepared by suspension polymerization are shown in the dry state in Fig. 1a, in which their rough, but prevalently regular spherical shape can be seen. The same granules swollen in water (Fig. 1b) are larger and smoother. Most of the particles (88%) have a bead diameter, in the dry state, in the range 0.4 - 0.8 mm; this is evident from the particle size distribution curve (Fig. 2).

Figure 1. Microphotographs of spherical particles of poly(2-hydroxyethyl methacrylate): (a) in the dry state; (b) swollen in water. (× 8.5).

In water, the diameter of the bead increases quickly, i.e. within 10 min by 20%, which is the equilibrium value. If the monomeric mixture contains, in addition to 2-hydroxyethyl methacrylate, 4 wt.% methacrylic acid, a substantial increase in the bead size in the swollen state takes place, e.g. in the Na^{+} cycle it is by 110%.

The amount of sorbed water at a constant vapor pressure is virtually unaffected by temperature (Fig. 3). The calculated value of the diffusion coefficient is 4×10^{-6} cm^2/s. This order of magnitude corresponds to the diffusion coefficient of poly(2-hydroxyethyl methacrylate) synthesized earlier by the conventional method [2]. The porosity of spherical hydrogel particles reaches 60% in water, and the specific surface area amounts to 0.9 m^2/g in the dry state.

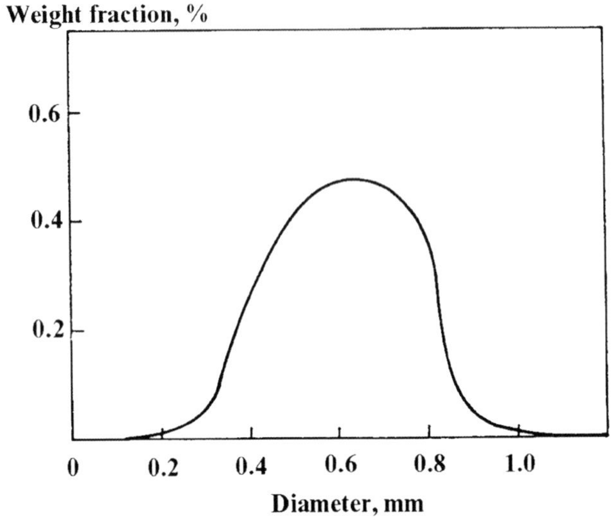

Figure 2. Differential curve of hydrogel particle size distribution.

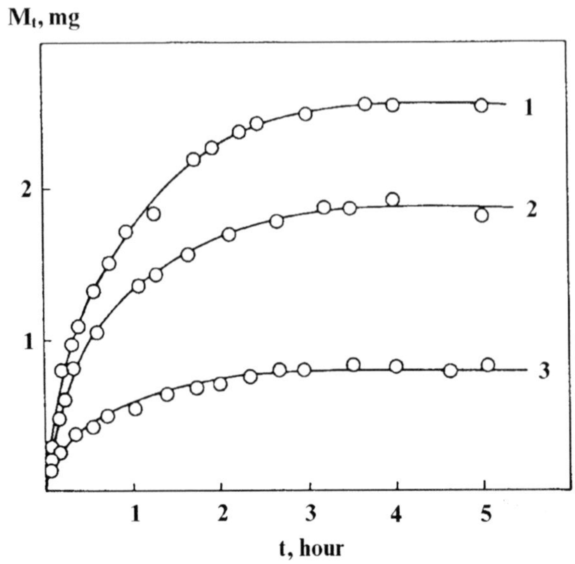

Figure 3. Kinetics of water vapor sorption by spherical hydrogel at various temperatures (1 - 30°C; 2 - 36°C; 3 - 44°C) and constant vapor pressure 3.5 kPa; M_t is the amount of sorted water in time t.

The diffusion and sorption characteristics of the spherical particles ensure their fast swelling in the blood stream and the expectant localization in the blood vessel. The swelling of poly(2-hydroxyethyl methacrylate) beads ensures reliable stability of the occlusive effect in the vascular lumen and excludes their possible migration as a result of blood pressure, on one hand, and the formation of firm thrombotic masses, on the other.

1.3.3 Haematological Testing

Blood analyses, or analyses of changes in the number of blood elements due to low-molecular weight compounds released from the hydrogel during its washing, can be advantageously employed in the determination of the quality of purification of artificial emboli prior to their use.

After washing-out the hydrogel for 1 day, a decrease in the number of erythocytes caused by their haemolysis, and of thrombocytes due to aggregation of the latter is found (Table 1). At the same time, one can perceive cytolysis and the formation of aggregates of leucocytes accompanied by a decrease in their number. Values in Table 1 show that after washing-out the hydrogel for 5 days there is no further decrease in the number of erythrocytes and leucocytes; hence, the low-molecular weight compound content is nontoxic, amounting to 10^{-5} g per 1 g of polymer at most, according to spectrophotometric measurements. Below this value all blood cells in contact with the extract retain their natural character. In this way the analysis of the blood morphology indicates the nontoxic amount of residual impurities in the polymer.

Table 1. Effect of the Length of Time of Washing-Out the Hydrogel on the Morphological Structure of Donor Blood.

Blood elements	Number of blood elements in 1 *l*		
	Donor blood	Contact between donor blood and water extract after washing-out the hydrogel for:	
		1 day	5 days
Erythrocytes ($\times 10^2$)	3.85 ± 0.55	2.85 ± 0.23	3.51 ± 0.40
Leucocytes ($\times 10^9$)	5.60 ± 0.50	2.85 ± 0.37	5.10 ± 0.30
Thrombocytes ($\times 10^9$)	5.20 ± 1.30	5.10 ± 0.20	7.50 ± 0.30

If the hydrogel is first washed and the content of low-molecular weight compounds in it drops below the toxic limit (10^{-5} g/g), only thrombocytopenia can be observed in the contact of blood with the subsequent extract; this is accompanied by the formation of thrombocytic aggregates (8 - 10 aggregates per 1000 erythrocytes). Each aggregate contains 20 - 30 thrombocytes. This is caused by the effect of the residual monomer and of the other low-molecular weight compounds inducing aggregation of the thrombocytes. Adhesion of the latter is the first step in the process of thrombus formation.

When the amount of impurities in the hydrogel exceeds 10^{-5} g/g, contact between blood and the hydrogel extract leads to morphological changes in erythrocytes (decrease in their size and changes in shape i.e. micro- anisocytosis). One can see fragments of erythrocytes and their fragmentation. The haemoglobin level decreases by 40 - 50%. The haemolysis level increases to 50 - 60 mg %. At the same time, there is a great change in the morphology of white blood cells - leucocyte artefacts (10 per 100 cells), pathological shape of neutrophiles and shrinkage of the nuclei. The number of thrombocytes decreases 3 - 5 times compared with the standard.

The physical character of the surface (roughness) and of the interior of hydrogel particles makes sorption of blood proteins and cell elements possible. In this respect the hydrogel can be regarded as a thrombogenic material. At the same time, one can see selective adsorption of fibrinogen. This facilitates the following adhesion and aggregation of thrombocytes onto the hydrogel surface and rigid fixation in the place of occlusion.

Both processes which appear during implantation of the hydrogel, i.e. both desorption of impurities and sorption of proteins help to bring about quick blood coagulation and thrombus formation on the particle surface, i.e. support the process of endovascular embolization. Spontaneous aggregation of thrombocytes plays an important role particularly immediately after embolization.

1.3.4 Histomorphological Investigation

The last medico-biological test consisted of a histological investigation of the material which had been implanted in the subcutaneous cellular tissue or blood vessel of the test animal for variously long periods of time. Two weeks after the implantation, the hydrogel bead, subcutaneously localized in the rabbit, is surrounded by a fibrous capsule without any dystrophic and (or) inflammatory changes (Fig. 4). The hydrogel is in an early stage of organization. Individual fibroblasts fill the pores of the hydrogel embolus, mainly in the peripheral domains of the bead.

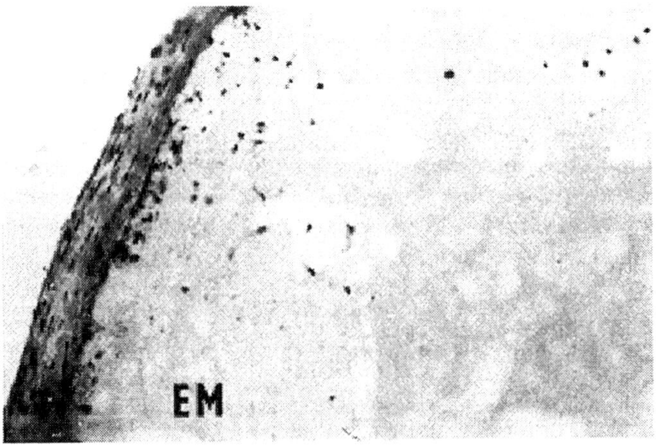

Figure 4. Histological section of subcutaneously localized spherical hydrogel 2 weeks after implantation. EM, embolus material occupying the largest part: fibrous capsule is on the left-hand side. Stained with haematoxylin andeosin. (× 106).

Five weeks after the implantation, the number of fibroblasts inside the pores of the peripheral domains of the bead is increased (Fig. 5). Thin bunches of collagen fibers (resulting from the biosynthesis of fibroblasts) can be seen in the pores. There are also some blood capillaries.

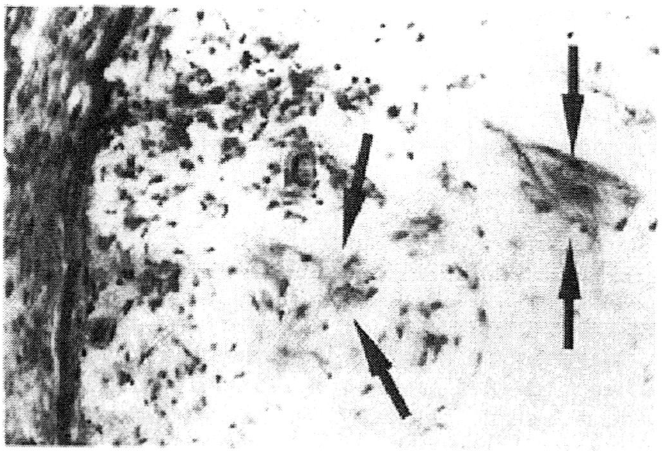

Figure 5. Histological section of subcutaneously localized spherical hydrogel 5 weeks after implantation. C, capillaries, arrows indicate collagen fibers. Stained after van Gieson. (× 166).

An earlier investigation [10] of the tolerance of heterogeneous poly(2-hydroxyethyl methacrylate) gels in the living tissue also confirmed that macroporous polymers were encapsulated with a capsule of collagen fibrous tissue, from which newly formed capillaries and numerous cells penetrated into the implants. The more porous the material, the broader the zone of cellulization. On the other hand, however, gels intergrown with the newly formed fibrous tissue is an inconvenient, as implants in plastic surgery, since calcification occurs, depending on the degree of porosity in heterogeneous macroporous polymers. At the same time, for the purpose of endovascular embolization, a highly porous structure formed by interconnected channels is an essential condition in the vascular lumen for anchoring the gel by the fibrous tissue.

Twenty-one months after implantation of a hydrogel bead into the bronchial artery, its wall had atrophied and sclerosed (Fig. 6). There were no inflammatory or destructive changes to the wall. The embolus material is seen to be divided by the accumulation of giant cells of foreign bodies (resorption) and by the streaks of fibrous tissue cells growing into the pores (organization). The recanalization of the vascular lumen is absent.

Consequently, investigation of the responses of tissue surrounding spherical hydrogels, after various periods of time from implantation, revealed that a thrombus had been formed on the surface of the embolus and inside the cavities between the hydrogel and the blood vessel immediately after the occlusion. This thrombus is replaced step by step by a connective tissue. All indistinct resorption due to the reaction of the giant cells and the partial organization of the hydrogel beads prevailed in the case of medium-long periods of time (e.g. 7 weeks) after implantation. These phenomena were more intense with decreasing bead size. Chronic inflammation and destructive-dystrophic changes (necrosis) of the tissue were absent. Resorption of hydrogel emboli does not cause recanalization or reopening of the vascular lumen, which is especially important in the clinical application of hydrogels to endovascular occlusion. The polymeric material did not display any mechanical destruction more than 1 year after implantation [11, 12].

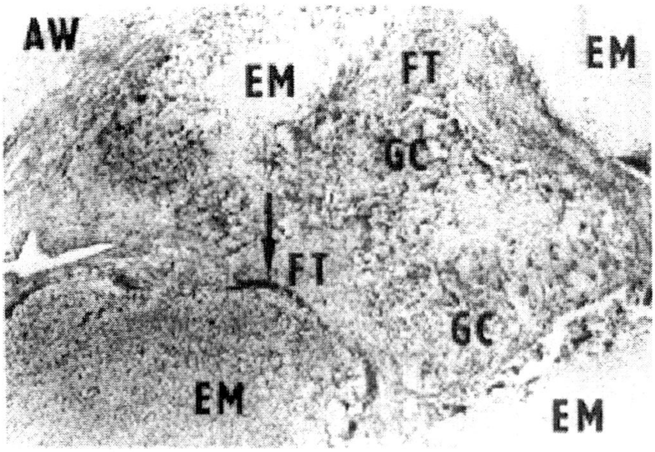

Figure 6. Histological section of the bronchial artery with spherical hydrogel 21 months after implantation. AW - artery wall; EM - embolus material; GC - giant cell; FT - fibrous tissue. Stained with haematoxylin and eosin (× 42).

References

[1] O. Wichterle, Hydrogels, in *Encyclopedia of Polymer Science and Technology*, **15**, Supplement, Interscience, J. Wiley & Sons, New York, 1971, 273-291.

[2] M.F. Refojo, "Permeation of Water Through Some Hydrogels", *J. Appl. Polym. Sci.*, **9**, 1965, 3417.

[3] K.H. Lee, J.G. Jee, M.S. Jhon, T. Ree, "Solute Transport Through Crosslinked Poly(2-hydroxyethyl methacrylate) Membrane", *J. Bioengineering*, **2**, 1978, 269.

[4] L. Sprincl, J. Vacik, J. Kopecek, "Biological Tolerance of Ionogenic Hydrophilic Gels", *J. Biomed. Mater. Res.*, **7**, 1973, 123.

[5] D.F. Williams, R. Roaf, *Implants in Surgery*, W.B. Saunders, London, 1973, p. 279.

[6] H. Tanzawa, in *Medical Use of Polymers*, Ed. S. Manabu, (Translation into Russian), Medicina, Moscow, 1981, p. 108.

[7] M. Stol., P. Lopour, I. Cifkova, "Process of Preparation of Sponge-like Synthetic Hydrogels with Improved Mechanical and Biological Properties", *Czech. Pat. Appl.*, 1983, 3372-3383.

[8] Yu.V. Moiseev, G.E. Zaikov, *The Chemical Resistance of Polymers*, Khimia, Moscow, 1979, p. 104. (in Russian).

[9] D. Horak, Z. Pelzbauer, M. Bleha, M. Ilavsky, F. Svec., J. Kalal, "Effect of Composition of Polymerization Feed on Morphology and Some Physical Properties of Macroporous Suspension Copolymers Glycidyl Methacrylate-ethylene Dimethacrylate", *J. Appl. Polym. Sci.*, **26**, 1981, 411.

[10] L. Sprincl, J. Kopecek, D. Lim, "Effect of the Structure of the Poly(glycol monomethacrylate) Gel on the Calcification of Implants", *Calc. Tiss. Res.*, **13**, 1973, 63.

[11] G.E. Zaikov, A.L. Buchachenko, V.B. Ivanov, in *Aging of polymers, polymer blends and polymer composites*, Nova Science Publ., New York, 2002, vol. 1, pp. 195.

[12] G.E. Zaikov, A.L. Buchachenko, V.B. Ivanov, in *Aging of polymers, polymer blends and polymer composites*, Nova Science Publ., New York, 2002, vol. 2, pp. 205.

2. Clinical Use of Spherical Particles

2.1 Introduction

In the preceding part of this Chapter and the associated paper [1] we reported the synthesis as well as physicochemical and medico-biological properties of biocompatible spherical hydrogel particles based on poly(2-hydroxyethyl methacrylate). These particles (called artificial emboli) were prepared for the occlusion of blood vessels, especially in pulmonary hemorrhage. Pulmonary hemorrhage (bleeding and hemoptysis) is one of the most frequent causes for seeking medical assistance; at the same time, it is one of the most serious complications which affects considerably the course of the disease and the prognosis of both, the working ability and life itself. Pulmonary hemorrhage is a phenomenon accompanying many lung and heart diseases. According to various authors, 7 - 15% patients admitted to hospital for chest diseases suffer from hemoptysis or hemorrhage [2, 3]. Pulmonary hemorrhage is the main cause underlying increased death, which in the case of strong hemorrhage reaches an average of 20% of the total number of patients.

In those cases where the surgical procedure used to stop bleeding is rather risky for the patient's life due to his/ her serious health state, or to the extensive character of the process, hemorrhage can be stopped by introducing artificial emboli into the blood stream. A successful course of such treatment is considerably dependent on the properties of emboli chosen for occlusion. Up to now, materials used for this purpose were randomly selected and differed diametrically in their chemical structure and properties. Gelatine sponge emboli are those most widely used [4]. Several other materials are available, compressible as well as non-compressible. Muscle emboli, as well as autogenous blood clots have been used as compressible emboli, but have the disadvantage of being rapidly dissolved. Intravascular tissue adhesives constitute another category of embolic material [5]. These plastic materials, polymerizing after injection into the blood stream, are difficult to handle and so far this has limited their widespread use. The shortcomings of haemostatic sponges based on gelatin are that within 2 - 4 weeks they undergo lysis, and revascularization commonly occurs. It is recommended that, when Spongostan is used as an embolic material preoperatively in order to decrease the bleeding at a subsequent operation, the operation should be performed within the first few days following embolization, before revascularization has taken place. Some other materials, such as china or steel beads, spherical microparticles made from polystyrene, suspension of barium sulfate, polyurethane sponge, or poly(methyl methacrylate) also did not meet the surgeon's requirements [1].

At present, special embolization in surgical practice is implemented with Gianturco coils in combination with velour felt from teflon [6]. However, their use is also associated with the following disadvantages. Firstly, the manufacture of the coils, which must be prepared for each patient individually immediately before the operation, is not a simple procedure and secondly, the transcatheter introduction is not easy to use. In addition, application of the coil may produce aneurysm of the occluded blood vessel [7].

Consequently, development of biocompatible emboli suited to the occlusion of blood vessels and studies of the indications and counter-indications for their clinical application are of great practical importance. Devascularization by embolization is indicated in tumors and vascular malformations either as a complement to surgery or as a single occlusive measure. Facilitation of surgery by preoperative embolization, or the occasional transformation of an inoperable lesion into one that is surgically accessible, is of practical interest, e.g. in therapeutic transcatheter embolization in the region of the external carotid artery [8]. Embolization offers the possibility of occluding arteries far from the periphery, by selecting emboli of a suitable size. The ideal would be to begin with small emboli to occlude the peripheral branches and then to continue with larger emboli in order to achieve a more proximal occlusion of the main feeding arteries [9].

In this part of the Chapter some aspects of the preparation of spherical emboli from poly(2-hydroxyethyl methacrylate) and the clinical results obtained from their application in termination or reduction of hemorrhage are discussed.

2.2 Experimental Section

2.2.1 Materials

The procedure used in the preparation of hydrogel particles and their classification according to size, has been described earlier [1]. A narrow particle size distribution according to the diameter of the vessels to be embolized (see Results and Discussion) is always required by the surgeons. Small particle size is employed in cases where at first small vessels are to be embolized and then larger ones, as a large embolus could prevent embolization of small vessels. Polydispersity is therefore undesirable. Particle size, in the dry and swollen state, was determined on the same particles as those used by surgeons in clinical applications.

2.2.2 Methods

The rate (virtually kinetics) of swelling of spherical particles in blood was determined according to the following method. Samples of spherical emboli on a support made from unwoven linen cloth, 1 mm thick, were placed on a perforated plate (Fig. 2.1) (25 perforations in an area 3.5 cm in diameter; perforation diameter, 3 mm) which was placed in a vessel thermostatted to 37°C and filled with citrated blood. An increase in the size of each particle was measured from the side in a horizontal GM 3 microscope equipped with an ocular micrometer. To obtain a statistical viable change in particle size at least 100 particles were measured.

Spherical particles were purified for medical use [1], in the container sterilized in vapor at 120°C and 11 Pa, for 8 min; the quality of sterilization of each batch was bacteriologically tested. The following method was used for the introduction of sterile spherical particles: firstly, 10 - 15 ml of 0.25% Novocaine solution was infused into the vascular lumen through a catheter with 1 mm i.d., then 20 ml of contrast medium (76% solution of Verografin) was injected, so that the final correct positioning of the catheter in the branch to be embolized could be confirmed. The blood vessels were then embolized by injecting a suspension of emboli, particle size 0.4 - 0.6 and 0.6 - 0.8 mm, in saline, through a catheter. The number of emboli needed for occlusion of a main branch vessel varied depending on the type of lesion and size of the vessel. The whole amount of the saline was about 200 g. The end-result was always documented angiographically before withdrawal of the catheter from the artery. In the presence of a pathological hyper-vascularization of large arterio-venous anastosomes in the focus, larger emboli (0.8 mm) were injected first, followed by smaller ones (0.4 - 0.6 mm). In such cases the particles were classified in advance using stainless steel templates with apertures 0.4, 0.5, 0.6, 0.7 and 0.8mm [1]. Used particles were already graded by sieving into fractions 0.4 - 0.6 and 0.6 - 0.8 mm. In the case of pathological hyper-vascularization of small blood vessels without anastosomes, the occlusion began with small emboli which were replaced step by step with larger ones, until the vessels were closed from the periphery to the centre. The degree and character of occlusion were continuously checked by angiography.

Figure 2.1. Apparatus for the determination of rate of swelling of porous materials: 1 - sample; 2 - perforated plate; 3 - thermostatted vessel; 4 - float.

2.3 Results and Discussion

One of the main requirements of spherical particles used in embolization is a narrow particle size distribution. The presence of very small particles in the overall amount of emboli may lead to a serious complication, i.e. tissue necrosis due to the intrusion of emboli into blood vessels not intended for embolization. The presence of large particles means the danger of the catheter being occluded and this would necessitate its exchange during the course of the operation. This raises the degree of traumatization of the tissues and extends the operation time.

Another very important parameter in the selection of emboli for the occlusion of blood vessels consists in an ability to increase their volume in contact with blood. This is of extreme importance if the emboli are to be quickly and reliably fixed in the blood stream. Fig. 2.2 shows the time dependence of the swelling of the emboli in various liquid media. The increase in size of hydrogel particles in blood is significantly larger than in water or saline (by 10%). This facilitates a subsequent quick fixation of the embolus in the blood stream where it arrives from the catheter in saline, since - as shown in Fig. 2.2 - equilibrium swelling is reached within 2 - 3 min after contact with blood. The increase in the particle size in blood compared with water or saline may be explained by the adsorption of blood proteins and aggregation of thrombocytes on the particle surface and by coagulation of blood.

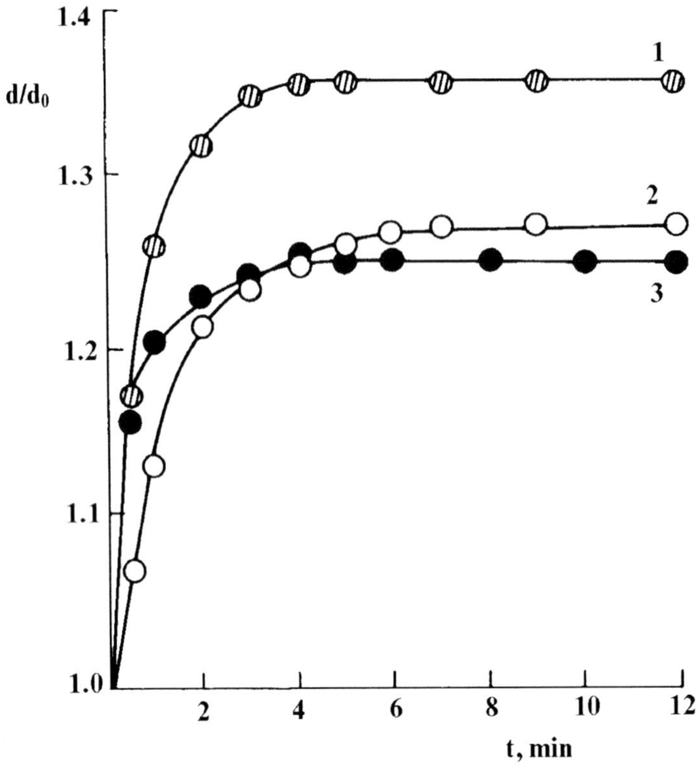

Figure 2.2. Time dependence of the relative particle size $\left(\frac{d}{d_0}\right)$ in various media: 1 - citrated blood; 2 - saline; 3 - water: d_0 is the particle diameter in the dry state; d the particle diameter in the swollen state.

Spherical emboli made from poly(2-hydroxyethyl methacrylate) were used to stop hemorrhage in patients suffering from lung and heart diseases. Fig. 2.3 (a, b) shows arteriograms of bronchial blood vessels before and after the embolization of a patient, aged 22, with a diagnosis of chronic pneumonia and pulmonary abscess. During the previous 3 months he was suffering from pulmonary hemorrhage. Conservative treatment was ineffective. Arteriography showed hyperplasia of bronchial arteries with foci of hypervascularization in the abscess zone (Fig. 2.3a). The trunk of the artery is widened and twisted and the number of peripheral branches is abnormally large. One can see foci of hypervascularization of lung tissue. Transcatheter endovascular occlusion of bronchial arteries was carried out with hydrogel beads, 0.4 - 0.6 mm in size. The amount of wet beads measured by volume was 1.5 cm^3. After the embolization was completed all pathologically degenerated branches responsible for hemorrhage disappeared; only the beginning of the trunk of the bronchial artery is contrasted (Fig. 2.3b). Hemorrhage had stopped. No relapses were recorded. A control check-up after 4 months revealed that the amount of released sputum was much smaller. Sputum changed from purulent to mucous, which suggested a positive effect of occlusion on the inflammatory process in the lungs.

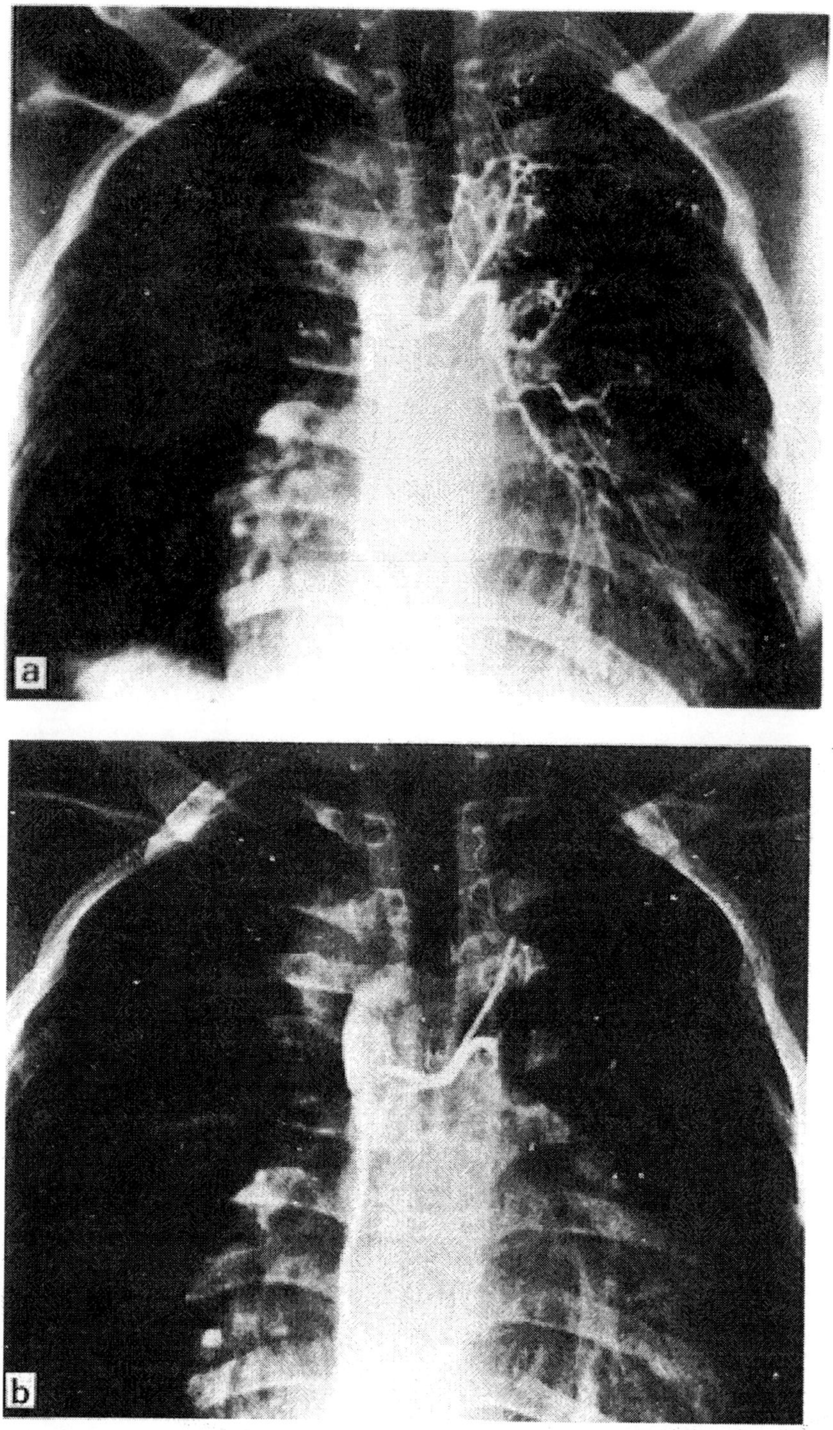

Figure 2.3. An arteriogram of bronchial blood vessels of a patient suffering from chronic pneumonia and pulmonary abscess before faf and after fbi embolization.

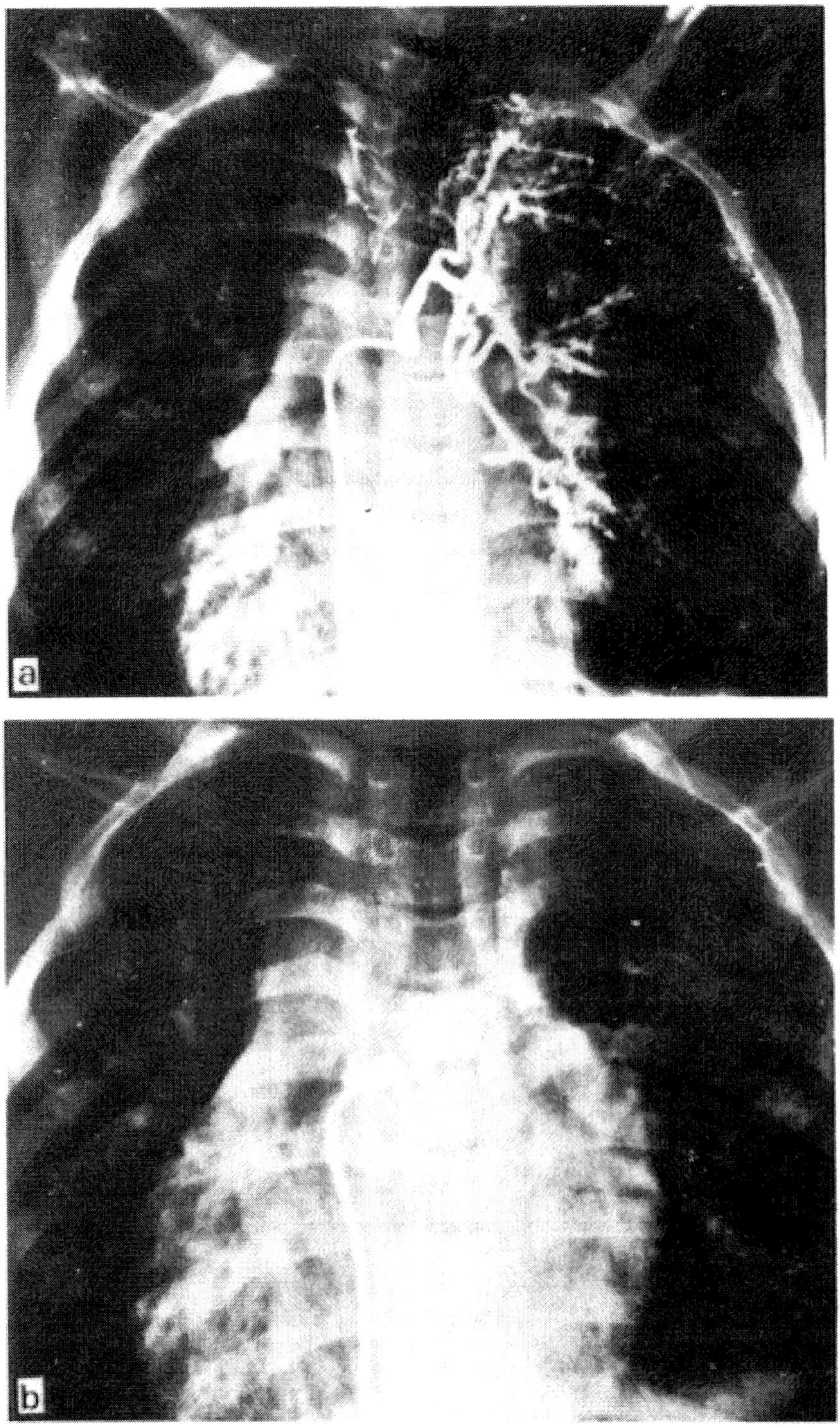

Figure 2.4. An arteriogram of the trunk of the intercostal artery of a patient suffering from Fallet's tetralogy and dextrocardia before (a) and after (b) embolization.

Emboli made from poly(2-hydroxyethyl methacrylate) were further used in the transcatheter occlusion of patients suffering from Fallot's tetralogy. In the pathogenesis of Fallot's tetralogy an important place is held by the hyperplasia of bronchial and other systemic pulmonary arteries. The hyperplasia of collateral arteries brings about important complications, with pulmonary hemorrhage being the most frequent one. A conservative treatment of pulmonary hemorrhage in Fallot's tetralogy does not give the desired results, and the existing ways of operative treatment involve a considerable risk to the patient's life. Fig. 2.4 (a, b) shows arteriograms of bronchial arteries before and after the embolization of the patient, aged 14, with diagnosis of Fallot's tetralogy and dextrocardia. When hospitalized, the patient suffered from frequent hemoptysis. The blood volume in hemoptysis was 50 - 70 ml. Medical examination revealed (Fig. 2.4a) a considerably widened trunk of the intercostal artery which supplied blood to the left lung and to the third left intercostal space. Bronchial arteries were considerably twisted and formed loops. In the peripheral parts of the artery one can see foci of hypervascularization with bronchopulmonary shunts.

Transcatheter occlusion of the intercostal artery was achieved using poly(2-hydroxyethyl methacrylate) beads, 0.4 - 0.6 mm in diameter, in an amount corresponding to 2.5 ml. During the occlusion of the bronchial artery with the emboli, the bronchopulmonary anastosomes and the branched network of arteries which were the source of the hemorrhage were blocked. After successful embolization, the control angiography (Fig. 2.4b) showed that only that part of the basic trunk of the artery which does not contain emboli is contrasted; the blocked blood vessels do not contrast. Hemoptysis did not occur any more. The patient was released in a satisfactory state. A later check-up after 24 months confirmed the correctness of the treatment.

Hydrogel spherical particles were used in 187 patients in the occlusion of bronchial, renal, mesenteric, hypogastrica interna and other arteries in order to stop organic hemorrhage and to influence the course of the disease. Due to the large inner diameter of the anastosomes, large emboli, 0.8 mm in size, were applied to 50 of 175 patients suffering from artero-venous anastosomes in the focus of hypervascularization. An evident haemostatic and clinical effect could be seen in an immediate and further (1 - 2 years) observation.

In the cases of hemorrhage and hemoptysis the introduction of emboli resulted in an immediate haemostatic effect in 86.6% of cases. Unsatisfactory results were caused by incomplete occlusion of the main feeding arteries, by some special features of the disease and in 3 cases by the delicate state of the patient. Otherwise, even after a considerable time interval, the results of embolization ranged from good to excellent in 92% cases.

Experience provided by successful clinical applications of spherical emboli prepared from poly(2-hydroxyethyl methacrylate) and intended for the occlusion of blood vessels allows us to infer that this material, which can be obtained by employing a simple procedure, is nontoxic, perfectly biocompatible, and possesses properties which make it particularly suitable for stopping hemorrhage by the method of transcatheter embolization.

A forthcoming part will deal with the preparation and properties of X-ray contrast spherical hydrogel particles [10, 11].

References

[1] D. Horak, F. Svec, J. Kalal, K.Z. Gumargalieva, A.A. Adamyan, N.D. Skuba, M.I. Titova., N.V. Trostenyuk, "Hydrogels in Endovascular Embolization. I. Spherical Particles of Poly(2-hydroxyethyl methacrylate) and Their Medico-biological Properties", *Biomaterials*, **7**, 1986, 188-192.

[2] M.I. Kuzin, Yu. D. Volynskii, "Etiology and Pathogenesis of Pulmonary Hemorrhage", in *Endovascular Therapeutics and Surgery of Pulmonary Hemorrhage*, Vostochno-Sibirskaya Pravda, Irkutsk, 1981, 7-11 (in Russian).

[3] J.D. Wolfe, D.U. Simmons, "Hemoptysis Diagnosis and Management", *West. J. Med.*, **127**, 1977, 383-390.

[4] Kh. Rabkin, M.I. Perelman, Yu.V. Biryukov, J.N. Gotman, R.K. Ambrozaitis, "Experiment of Therapeutic Embolization of Bronchial Arteries in Pulmonary Hemorrhage and Hemoptysis", *Khirurgiya*, **12**, 1979, 8-13 (in Russian).

[5] J. Clarisse, G. Gozet, J.P. Cornil, M. Jomin, J.M. Delandsheer, E. Laine, Les Emboles Plastiques Fluides. Etude Experimentale et Rapport de Deux Cas Cliniques de Fistulas Arterioveineuses Trairtes, *J. Neuroradiol.*, **2**, 1975, 29-38.

[6] Kh. Rabkin, V.A. Klimanskii, L.N. Gotman, G.N. Zakharov, Embolization of Splenic Artery with Disease of Blood System, *Khirurgiya*, **2**, 1981, 53-55 (in Russian).

[7] L. Vlahos, S. Trakadas, P. Kehyas, G. Pontifex, Transcatheter "Arterial Embolization With Gianturco Steel Coils: A Complication", *Acta Urol. Belg.*, **51**, 1983, 256-260.

[8] J. Brismar, S. Cronquist, "Therapeutic Embolization in the External Carotid Artery Region", *Acta Radial. Diagn.*, **19**, 1978, 715-731.

[9] L. Picard, J.M. Andre, J. Roland, M. Sigiel, J. Montaut, J. Lepoire, L'embolisation dans les Malformations Vasculaires Meningo-cranio-cutanees Complexes, *J. Neuroradiol.*, **2**, 1975, 233-256.

[10] G.E. Zaikov, "*Polymers in medicine*", Nova Science Publ., New York, 1999, pp.165.

[11] G.E. Zaikov, A.L. Buchachenko, V.B. Ivanov, *Polymer aging at the cutting edge*, Nova Science Publ., New York, 2002, 176 pp.

3. Radiopaque Spherical Particles, Their Preparation and Properties

3.1 Introduction

In preceding communications in this series we have reported the preparation of spherical hydrogel particles, their properties [1] and clinical performance in vascular occlusion [2]. They were employed for the embolization of tumor arterial blood supply, either in inoperable tumors or to prevent hemorrhage during surgery. Embolization is used in the treatment of non-malignant and malignant neoplasmas, especially hemangiomas, arterio-venous anastosomes and vascular malformations, organ hemorrhage and hemoptysis, and the like. Soft and readily swellable spherical hydrogel microparticles have many advantages in clinical applications: they are perfectly biocompatible, the living tissue grows through them without inflammation or causing destructive dystrophic changes. The thrombus is formed immediately after the application of the embolus which is quickly fixed in the vascular lumen where it has a permanent haemostatic effect without recanalization. Last but not least, the biggest advantage of these particles is the simplicity of their introduction into the bloodstream by means of a syringe and catheter.

On the contrary, materials used so far in therapeutical practice present many shortcomings e.g., only a short-term occlusion effect, the necessity of a complicated operation, insufficient biocompatibility, and also lack of radiopacity. However, in order to keep embolization under control, permanent surveillance of the embolization process and implant's behavior in the post-operation period is imperative. Up to now, this process has been monitored only by angiography. Therefore, radiopaque emboli, which can be directly monitored by X-ray camera immediately after their introduction, as well as for a long period of time after implantation, represent a qualitatively higher category of material suitable for the embolization of blood vessels.

The first X-ray contrast materials available for the occlusion of an arterio-venous fistula were fluid plastic emboli containing radiopaque oil; however, certain toxicity, and difficult handling during influx [3, 4] limit their application. Recently, the gels for therapeutic embolization, containing X-ray contrast media, have been reported in the Patent literature. They are used to prevent hemorrhage or to produce ischaemia of malignant tumors. Celling agents are based on cellulose [5], gelatin [6], silicone rubber [6], acetylcellulose [7], or a copolymer of vinyl pyrrolidone and methyl methacrylate [7]. The latter two materials clog the catheter in cases where the addition of poly(ethylene glycol) is omitted. Tantalum powder or $Fe_2O_3{\cdot}6BaO$ is also recommended [7] as a contrast substance. However, by using these substances *in vivo*, the radiopaque additive is readily washed out from the embolus. This shortcoming can be overcome by particles with the radiopaque compound chemically bonded to the polymer chain. In this study we report the synthesis and properties of such spherical hydrogel particles.

3.2 Experimental

3.2.1 Materials

Monomers and Auxiliary Compounds

Chemicals were redistilled prior to use: ethylene dimethacrylate (Ugilor S.A., France) - b.p. 50°C/7 Pa, cyclohexanol (Lachema Brno) - b.p. 68°C/1.6 kPa, decahydronaphthalene (Lachema Brno), methacrylic acid (Ferak, Berlin), thionylchloride (Cambrian Chemicals, Great Britain), dioxan. 2,2'-Azo(bis-isobutyronitrile) (Ferak, Berlin) was recrystallized twice from ethanol. 2-Hydroxyethyl methacrylate (Leciva Modrany), medicinal purity, was used without further purification. The other chemicals used were reagent grade. The preparation of poly(1-dodecyl methacrylate) has been described in a previous communication [1].

3.2.2 Methods

Preparation of Particles Containing Barium Sulfate by Inverse Phase Suspension Polymerization

The 250 ml glass reactor equipped with an anchor-type stirrer was filled with 25 ml of dispersed phase containing 15 ml of a 20 % (w:w) aqueous solution of barium chloride, 9.3 ml 2-hydroxyethyl methacrylate, 0.3 ml ethylene dimethacrylate, 0.4 ml methacrylic acid, 0.1 g 2,2'-azo(bis-isobutyronitrile) and 75 ml of dispersing medium consisting of a 5 % (w:w) solution of poly(1-dodecyl methacrylate) (molecular weight 4×10^6) in decahydronaphthalene. The polymerization mixture was degassed by bubbling nitrogen through for 10 min. The polymerization proceeded under continuous stirring (400 rpm) at 70°C for 8 h. After cooling, the polymer was obtained as an agglomerate of particles difficult to process. It was separated and decanted by hexane and ether. Barium sulfate was generated by 5% sulfuric acid.

Preparation of Particles Containing Iopanoic Acid by Suspension Polymerization

25 ml dispersed phase containing 1.3 g iopanoic acid [β-(3-amino-2,4,6-triiodophenyl)-α-ethylpropionic acid - trade name Jopagnost Spofa] dissolved together with 0.1 g 2,2'-azo(bis-isobutyronitrile) in 7.5 ml 1-dodecanol (Fluka A.G.), 7.5 ml cyclohexanol, 9.8 ml 2-hydroxyethyl methacrylate and 0.2 ml ethylene dimethacrylate was placed in the reactor followed by 75 ml of a 1% aqueous poly(vinyl pyrrolidone) K-90 (Fluka A.G.) as the dispersing medium. Turbidity due to the Jopagnost precipitation occurred after the addition of the aqueous dispersing phase. After bubbling nitrogen through the polymerization proceeded at 70°C for 8 h under continuous stirring at 240 rpm. The resulting beads were separated, decanted by water, methanol, acetone and ether, and dried.

Preparation of Particles by a Modification of the Glycidyl Methacrylate - Ethylene Dimethacrylate Copolymer with 2-Aminoethylamide of 3-Amino-2,4,6-Triiodobenzoic Acid

The preparation of the sample spherical macroporous copolymer of glycidyl methacrylate (75%) and ethylene dimethacrylate (25%) has been described elsewhere [8]. 2-Aminoethylamide of 3-amino-2,4,6-triiodobenzoic acid was synthesized according to a Czechoslovak Patent [9] by reacting ethyl 3-aminobenzoate with ethylenediamine (Fluka A.G.) followed by iodination. The glycidyl methacrylate beads (150 - 180 mm) (0.1 g) were soaked with 6 ml dimethylformamide containing 0.6 g of 2-amino ethylamide 2 g of 3-amino-2,4,6-triiodobenzoic acid, and the suspension was shaken in a thermostated bath at 60°C for 260 h. The product was washed with dimethylformamide, acetone, ether, and dried.

The Acylation of 2-Hydroxyethyl Methacrylate-Ethylene Dimethacrylate Particles with 3-Acetylamino-2,4,6-Triiodobenzoyl Chloride

This was according to the following Scheme:

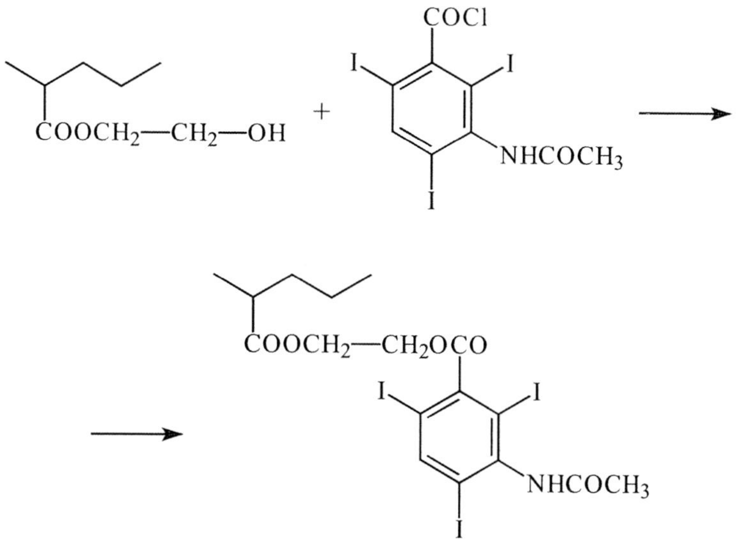

Table 3.1. Modification of 2-Hydroxyethyl Methacrylate-Ethylene Dimethacrylate Particles

No.	Molar ratio ROS/-OH	Molar ratio TEA/ROS	Concentration, mmol ROS/ml dioxan	Iodine content in product, wt.%
1	2	1.1	0.3	43.5
2	2	—	0.3	21.7
3	1	1.1	0.2	34.9
4	1	—	0.2	14.4
5	0.5	1.1	0.1	23.4
6	0.5	—	0.1	8.4
7	0.5	1.1	0.15	35.1

ROS, Radiopaque substance (3-acetylamino-2,4,6-triiodobenzoyl chloride); -OH, hydroxy groups of poly(2-hydroxyethyl methacrylate): size of sample beads 1 - 1.2 mm; TEA, triethylamine. See Experimental section for details.

The synthesis of sample hydrogel beads by the suspension radical copolymerization of 2-hydroxyethyl methacrylate (98%) with ethylene dimethacrylate (2%) has been described earlier [1]. 3-Acetylamino-2,4,6-triiodobenzoyl chloride was prepared [10] by reacting 3-acetylamino-2,4,6-triiodobenzoic acid (obtained from 2,4,6-triiodobenzoic acid, Leciva Modrany) with thionylchloride. In a typical reaction, 0.13 g (0.225 mmol) 3-acetylamino-2,4,6-triiodobenzoyl chloride was dissolved in 2 ml of dry dioxan with stirring at 50°C. After cooling to room temperature, 0.035 ml (0.25 mmol) triethylamine and 0.06 g (0.45 mmol -OH) hydrogel particles, 1 - 1.2 mm in size, were added. The mixture was shaken in a thermostated bath at 60°C for 150 h. The particles were then washed with dioxan and water until the OD_{230} of washing water falls below 0.03 (1 cm cell). Finally, the particles were washed 10 times each by acetone and ether, respectively (each portion contained ca. 15 ml of solvent) and then vacuum dried at room temperature. Polymer no. 5 (Table 3.1) was obtained according to this procedure. In the following experiments, the concentration and molar ratio of the radiopaque substance to the hydroxy groups were varied, or triethylamine was omitted (cf. Table 3.1).

Preparation of Particles by a Modification of the Terpolymer of 2-Hydroxyethyl Methacrylate, Ethylene Dimethacrylate and Methacrylic Acid with 3-Acetylamino-2,4,6-Triiodobenzoyl Chloride

The sample terpolymer beads were prepared as described above for the preparation of particles by inverse phase suspension polymerization, the only difference being that the addition of barium chloride was omitted. For modification of the final beads, the same procedure was adopted as for the modification of 2-hydroxyethyl methacrylate - ethylene dimethacrylate particles with 3-acetylamino-2,4,6-triiodobenzoyl chloride.

At least 100 particles, each of which was treated individually, were measured under the microscope to obtain a particle size in the dry and swollen state. Iodine was analyzed by Schoniger's method.

3.3 Results and Discussion

Several procedures for the preparation of radiopaque emboli have been tested. Firstly the suspension polymerization of 2-hydroxyethyl methacrylate in the inverse phase, where the dispersing phase consists of an organic liquid immiscible with water, and the aqueous solution of monomers constitutes the dispersed phase, has been exploited. In addition, barium chloride was dissolved in the aqueous phase and precipitated with sulfuric acid in the resulting product after the polymerization had been completed. However, this procedure failed because of the instability of the suspension due to the barium salt, and the resulting product in the form of an agglomerate became difficult to process.

Another original procedure, in which the radiopaque substance is precipitated in the hydrogel network, uses two water-soluble compounds. The hydrogel swollen in an aqueous solution of one of these compounds is immersed into a solution of the other compound [11]. Yet another method describes the preparation of radiopaque hydrogel by dissolution in a polymerization mixture of a radiopaque substance insoluble in water and its precipitation in the gel after swelling with water [12]. However, when using iopanoic acid [P-(3-amino-2,4,6-triiodophenyl)-α-ethylpropionic acid] dissolved in the polymerization mixture for the preparation of radiopaque beads according to the latter method, the precipitation occurred immediately after the aqueous dispersing phase had been added, i.e., before the polymerization had started, and the resulting particles lacked radiopacity. In contrast to all the methods reported here, where the radiopaque substance remains unhanded, we preferred the covalent attachment of the X-ray adsorbing group to the polymer, which prevented its release *in vivo* and facilitated the washing out of all unreacted low-molecular weight compounds after the preparation.

Modification of reactive glycidyl methacrylate - ethylene dimethacrylate beads using 2-aminoethylamide of 3-amino-2,4,6-triiodobenzoic acid was attempted, since these beads are available on the market. Although the radiopaque substance was used in a two-molar excess compared with the epoxy groups, only 13.4 % (w:w) of iodine was detected in the final product, and therefore the particles did not provide enough contrast. Moreover, the glycidyl methacrylate polymers did not prove to be perfectly biocompatible [13].

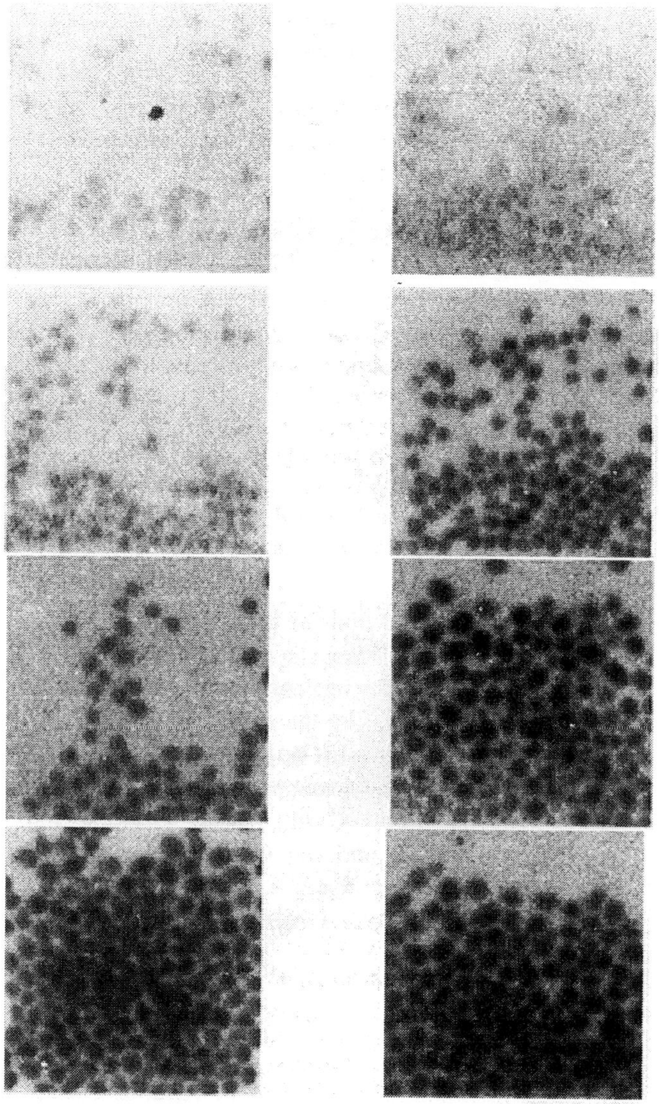

Figure 3.1. X-ray images of emboli with various iodine contents: (a) 8.3% I; (b) 8.9% I; (c) 14.4% I; (d) 27.6% I; (e) 23.4% I; (f) 34.9% I; (g) 35.7% I; (h) 43.5% I.

In the preparation of radiopaque particles based on poly(2-hydroxyethyl methacrylate) by modification with 3-acetylamino-2,4,6-triiodobenzoyl chloride, the amount of bonded radiopaque substance may be markedly affected by the molar ratio of the iodine-containing derivative to the hydroxy groups of hydrogel particles. Also it is markedly affected by the concentration of 3-acetylamino-2,4,6-triiodobenzoyl chloride in dioxan and by the addition of hydrogen chloride acceptor (triethylamine). The amount of the bonded radiopaque substance increases with increasing molar ratio of the iodine-containing derivative with respect to hydroxy groups and with its increasing concentration in the solvent, and is raised by the addition of triethylamine (Table 3.1). Particles containing 43.5 % (w:w) iodine were the maximum content product achieved. The amount of bonded radiopaque substance also depends on the size of the hydrogel particles at the start. Under otherwise identical conditions, more of the iodine-containing compound is bonded in smaller particles (Table 3.2) due to some kind of 'plugging' effect (or gradual filling of all pores accessible to the radiopaque compound); the surface layer which is penetrated by the iodinated molecule is limited. This layer represents a higher portion of the total bead mass in smaller particles, and therefore a higher average iodine content can be reached with smaller particles.

As shown in Fig. 3.1, the contrast of the emboli on the X-ray screen is directly related to the iodine content. The large volume of the molecule of triiodobenzoic acid derivative is one of the reasons why the size of the hydrogel particles increases with the degree of substitution (Table 3.3). Thus, e.g., at 43.5 % (w:w) iodine the particle diameter in the dry state is larger after the first and second drying cycles by 9% and 68%, respectively, compared with unmodified particles. It is known [14], that the porous structure of the original poly(2-hydroxyethyl methacrylate) particles collapses after two drying-swelling cycles and the particles become glassy. Modification by 3-acetylamino-2,4,6-triiodobenzoyl chloride partially suppressed this phenomenon, depending on the degree of substitution. At 43.5 % (w:w) iodine the radiopaque substance attached to the particles prevents the shrinking of the sample swollen beads during drying. Originally in the dry non-modified beads polymer chains were coiled. Upon swelling, these beads show a 20% increase in diameter and are fixed by acylation, because bulky iodine-containing substituents prevent the original compact (coiled) arrange ment of polymer chains after drying. After transfer to water there are no more changes in the size of the modified particles i.e. the beads do not swell. However, beads with a lower content of the radiopaque substance still partially retain their swelling ability. Thus, e.g. beads containing 10 % (w:w) iodine increase their diameter in water by 14%. Moreover, Table 3.3 shows the tendency to shrinking, i.e. to a decrease in the size of the dry particles after two cycles of drying and swelling which is less pronounced at higher iodine contents. Highly iodinated particles are harder, more brittle and less hydrophilic.

Table 3.2. The Size Effect of 2-Hydroxyethyl Methacrylate -Ethyiene Dimethacrylate Particles on their Acylation by 3-Acetylamino-2,4,6-Triiodobenzoyl Chloride (Given as the Iodine Content).

Fraction with particle sixe, mm	Iodine content in product, % (w:w)
0.6 - 0.8	30.6
0.8 - 1.0	29.3
1.0 - 1.2	28.7
1.2 - 1.5	27.9

Reaction conditions: molar ratio ROS/-OH = 0.5; TEA/ROS = 1.1; concentration is 0.12 mmol ROS/ml dioxan. For meaning of abbreviations see Table 3.1.

The low swelling of the particles impedes medical applications. The size of the emboli should facilitate the passage through the catheter, and at the same time they should swell in the bloodstream so as to be well fixed in the blood vessel lumen. It is known [15] that the connective tissue grows much more easily through high swelling porous particles.

The swelling of particles could be aided by copolymerization using several percent of methacrylic acid. Modified particles of such terpolymer in the Na^+ form containing 9.2 or 23.5 % (w:w) iodine increase in size in an isotonic solution of sodium chloride by 33% and

16%, respectively. Preliminary tests have shown, however, that already a small amount (4 vol.%) of methacrylic acid in hydrogel particles leads to blood haemolysis.

Thus, it can be seen that in practice a compromise should be sought between particles with a high iodine content which can be roentgenologically perfectly monitored and those with a low iodine content which swell well. Moreover, it is necessary for medical applications that emboli should also preserve a sufficient mechanical stability. These requirements are best met by particles with a 25 - 30 % (w:w) of iodine which can be adequately monitored when being introduced into the blood vessels and which also enable the checking of their performance in viva without the necessity for angiography.

Table 3.3 Effect of Iodine Content on the Size of Radiopaque Particles based on Poly(2-Hydroxyethyl Methacrylate) in the Dry, Swollen in Water, and the Dried Again States.

Iodine content,	Size of particles, mm		
% (w:w)	Dry	Swollen	Dried again
8.9	0.66	0.76	0.62
14.4	0.75	0.81	0.73
21.7	0.77	0.82	0.76
23.4	0.84	0.89	0.83
35.1	1.03	1.07	1.02
43.5	1.11	1.11	1.10

The size of poly(2-hydroxyethyl methacrylate) sample particles before start of reaction: in the dry state 1.02; swollen in water 1.23; and the dried again state 0.66 mm.

References

[1] D. Horak, F. Svec, J. Kalal, K.Z. Gumargalieva, A.A. Adamyan, N.D. Skuba, M.I. Titova, N.V. Trostenyuk, "Hydrogels in Endovascular Embolization. I. Spherical Particles of Poly(2-hydroxyethyl methacrylate) and Their Medico-biological Properties", *Biomaterials*, **7**, 1986, 188-192.

[2] D. Horak, F. Svec, J. Kalal, A.A. Adamyan, Yu.D. Volynskii, O.S. Voronkova, L.S. Kokov, K.Z. Gumargalieva, "Hydrogels in Endovascular Embolization. II. Clinical Use of Spherical Particles", *Biomaterials*, **7**, 1986, 467-470.

[3] J. Clarisse, G. Gozet, J.P. Cornil, M. Jorum, J.M. Delandsheer, E. Laine, Les Emboles Plastiques Fluides. Etude Experimentale et Rapport de Deux Cas Cliniques de Fistules Arterioveineuses Traitees, *J. Neuroradiol.*, **2**, 1975, 29-38.

[4] R.E. Krall, *Radiopaque Cyanoacrylate Compositions*, Europ. Patent Appl., No. 50457, 1982.

[5] P. Trampont, P. Madoule, B. Bonnemain, F. Puisieux, *Gel for Therapeutic Embolization*, Fr. Demande, No. 2 548 902, 1985.

[6] L. Lazar, K. Palossy-Becker, J. Naggy, E. Pasztor, "*Pharmaceutically Acceptable Silicone Rubber and Therapeutical Set and Their Use for Surgical Embolization*", Can. Patent No. 1 182 124, 1985.

[7] V. D. Solodovnik, T.I. Solodkaya, M.T. Litvinova, T.A. Meshkova, A.V. Usova, A.B. Davydov, "*Composition for the Embolization of Blood Vessels/Radiopaques*", Brit. Patent No. 2144327, 1985.

[8] D. Horak, Z. Pelbauer, M. Biota, M. Ilavsky, F. Svec, J. Kalal, "Reactive Polymers. XXXII. Effect of Composition of Polymerization Feed on Morphology and Some Physical Properties of Macroporous Suspension Copolymers Glycidyl Methacrylate - Ethylene Dimethacrylate", *J. Appl. Polym. Sci.*, **26**, 1981, 411-421.

[9] J. Drobnik, J. Kalal, M. Metalova, F. Rypdcek, "*Macromolecular X-ray Contrast Substance and Method of Its Preparation*", Czech. Patent No. 207858, 1984.

[10] H. Priewe, R. Rutkowski, K. Pirner, K. Junkmann, "Derivate der 2,4,6-triiod-3-amino-benzoesaure", *Chem. Ber.*, **87**, 1954, 651-658.

[11] N.V. Trostenyuk, K.Z. Gumargalieva, Yu.V. Moiseev, P. Lopour, J. Sulc, J. Vacik, J. Kalal, *The Method of Preparation of Hydrogels with X-ray Contrast Properties*, Czech. Patent Appl. No. 231700, 1984.

[12] P. Lopour, J. Sulc, J. Vacik, J. Kalal, T.T. Daurova, O.S. Voronkova, N.D. Skuba, *The Method of Preparation of X-ray Contrast Hydrogel*, Czech. Patent Appl. No. 231699, 1984.

[13] K. Filip, *PhD Thesis*, Institute of Clinical and Experimental Medicine, Prague, 1986.

[14] J. Hasa, *PhD Thesis*, Institute of Macromolecular Chemistry, Prague, 1966.

[15] L. Sprincl, J. Kopecek, D. Lin, Effect of Porosity of Heterogeneous Poly(glycol monomethacrylate) Gels on the Healing-in of Test Implants, *J. Biomed. Mater. Res.*, **5**, 1971, 447-458.

[16] E.Ya. Davydov, G.B. Pariiskii, A.P. Vorotnikov, G.E. Zaikov, Peculiarities of solid phase reactions, *John Wiley and Sons, Chichester (UK),* 1998, 150 pp.

[17] G.V. Kozlov, G.E. Zaikov, *Fractal analysis of polymers*, Nova Science Publ., New York, 2001, 244 pp.

4. Morphological Foundation of Hydrogel Use for Vascular Occlusion

4.1 Introduction

In clinical practice one can often face such situations, when conventional methods of hemorrhage control are not feasible or they require complicated surgical procedures threatening the patient's life [1]. Bronchial hemorrhage is the most vivid example in this respect, as it is conditioned by corrosion of the blood vessel wall. In such cases the safest and at the same time the most reliable method of hemorrhage control is endovascular occlusion by means of injection of different artificial emboli into the vessels [2]. Embolization is also needed in vascular surgery as a primary condition in the preoperative preparation of patients before excision of vascular tumors. In those cases when due to localization of a vascular tumor or developmental anomaly of vessels their removal is impossible, endovascular occlusion is the only possible way of treatment. The possibility of successful therapeutic endovascular occlusion depends upon the correct choice of the material used for embolus production. At present, the question of selection of the material for embolization purposes has not been solved definitively [3]. The aim of the present study is to determine the character of hydrogel embolus interaction with blood, vascular wall and with body soft tissues.

4.2 Materials and Methods

The material used in this study was:

(1) femoral arteries of rabbits (10 animals) occluded with hydrogel emboli;
(2) soft periembolar tissues, the subcutaneously implanted emboli to rabbits (10 animals);
(3) blood vessels and renal tissues of patients with renal tumors after renal artery occlusion with hydrogel emboli (2 patients);
(4) bronchial arteries and pulmonary tissues of patients suffering from chronic nonspecific pulmonary diseases (21 patients).

Materials were taken at different times after embolization, from a few days to 2 years. The main method of investigation was histological. In the early period (from 1 to 7 days after the embolus injection) microscopic studies of the histological specimens revealed the microporous structure of the hydrogel. Its pores were identified by the presence in them of plasma proteins or polymorphonuclear neutrophils. The hydrogel substance itself was not stained lay histological dyes. The thrombotic masses are observed in sites of vascular lumen, that are not filled with emboli. Thrombosis of the vascular lumen also occurs involving a small distance of the vessel distally and proximally to the emboli localization, as shown in Fig. 4.1.

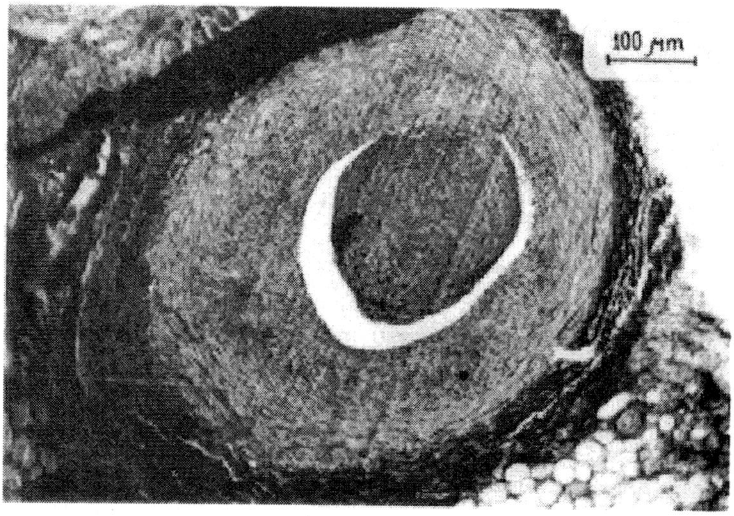

Figure 4.1. The cross section of the femoral artery 0.5 cm proximal to the embolus. Thrombus is in the vascular lumen, 1 month after the implantation. HE staining.

The vascular wall in the site of embolus localization looked thinned and distended in all observations. It is explained by the fact that hydrogel has a property to swell directly in the vascular lumen or in the depth of body tissues. The swelling of hydrogel emboli is accompanied by a significant increase of their volume, which ensures stability of the occlusive effect and excludes their possible migration as a result of blood pressure. Stability of the obturating effect is ensured in future by the formation of a firm aggregation of emboli and thrombotic masses in the interspaces between them. In neither of the observations made on the vascular wall, coming in contact with hydrogel, destructive-dystrophic changes, or inflammatory changes were noted (Fig. 4.2). It is indicated by the absence of injuring and irritating effects of this material upon viable tissues. Similar results were received at the investigation of soft tissues adjoining to the subcutaneously implanted hydrogel emboli: neither destruction, nor inflammatory reaction was noticed in them. The implanted material was gradually incapsulated. In studying the experimental material it was determined that in the late follow-up period (490 days) the surface layers of emboli were grown through with a connective tissue; connective cells (fibroblasts) were determined in the hydrogel pores as well as collagen fibers synthesized by them (Fig. 4.3).

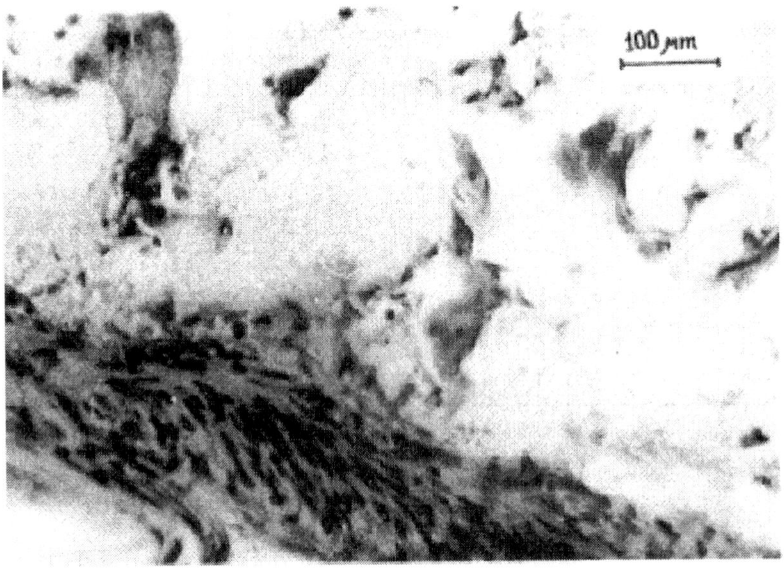

Figure 4.2. The hydrogel in the vascular lumen, 1 month after the implantation. The vascular wall is not changed. The hydrogel in the early stage of resorption and organization. Stained after van Gieson.

A small number of foreign body giant cells was noted in the superficial layers of emboli. The presence of these cells indicated an insignificant embolus resorption. The phenomena of embolus organization and resorption were more marked, when the emboli were subcutaneously implanted, than when they were placed directly into the vascular lumen. In the late time implantation (490 days), a partial embolus destruction of purely mechanical character occurs. It explains the presence of chipped off hydrogel particles in lymphatic fissures close to the autocutaneously implanted emboli. Thus, the experimental studies have shown, that hydrogel is a material with such properties that allow to recommend its use for blood vessel occlusion. Such features are: absence of injuring or irritating effect on tissues, capability to enhance thrombogenesis, while in contact with blood, the capability to provide a stable obturating effect. The results got in the experimental work allowed us to use hydrogel for blood vessel occlusion in clinical conditions. The successful endovascular occlusion with hydrogel emboli was performed on 110 patients in the Vishnevsky Surgery Institute. The data received on the operational material are essentially the same as in the experiment (Fig. 4.4).

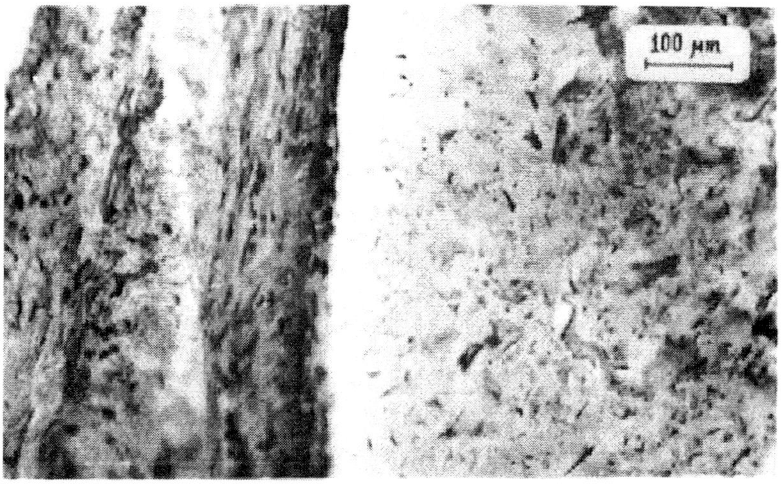

Figure 4.3. Fibrocytes and collagen fibers are within the pores of the hydrogel embolus, 490 days after the occlusion. Destructive-dystrophic changes or inflammatory changes are not observed. HE staining.

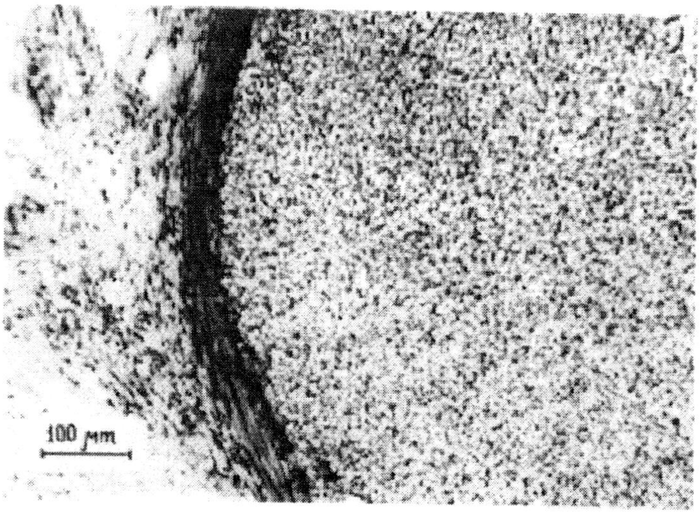

Figure 4.4. The lumen of a small bronchial artery filled up by a hydrogel embolus. The wall is thinned, 2 days after endovascular occlusion. HE staining.

Neither inflammatory reaction, nor any other destructive changes were noted in the wall of bronchial arteries in the hydrogel embolus localization, both in the early and in the late periods [4], (Fig. 4.5). Necrotic and destructive areas of pulmonary tissue were not observed in the bronchial wall and pulmonary parenchyma. The difference is only in the degree of the resorptive giant-cell reaction, when hydrogel was used for control of pulmonary hemorrhage in patients with chronic non-specific pulmonary diseases. Endovascular occlusion of bronchial arteries was performed in these patients. The giant-cell response in these observations was significant already a month after the embolization of bronchial arteries. Then it was increasing, so that to the 20th month some smallest hydrogel emboli were completely dissolved, and the vascular lumen was filled with the giant cells of foreign bodies or was overgrown with connective tissue.

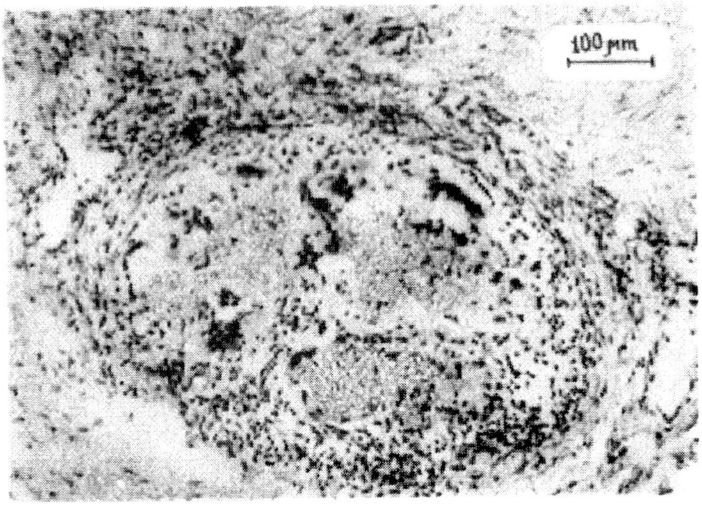

Figure 4.5. The cross section of a bronchial artery, 635 days after occlusion by the emboli. The wall of the artery is not infiltrated with inflammatory cells. HE staining.

We are inclined to account this peculiarity for the fact that, here the emboli were more accessible for hematogenous inflow of macrophagous elements. Such accessibility is due to abundant vascularization of the bronchial wall and presence of anastosomes at the precapillary level between the bronchial artery system and pulmonary artery system. When studying the surgical patients, to whom endovascular occlusion of bronchial arteries had been performed, it was possible to follow the character of rearrangement of the vascular net not only in the embolized zone, but distal and proximal to embolus localization. This process involved vacate hyperplasia of the intima, reorientation of smooth muscle cells from the circular to longitudinal direction, that in the long run resulted in the acute constriction of the blood vessel lumen. The reason for such rearrangement was the reduction of blood flow after endovascular occlusion. The hemostasis was obtained after endovascular occlusion as a result of the decrease of blood filling in the bronchial wall. At the same time fading of the inflammatory process with the reduction of the amount of sputum secretion occurred. Frequently the patients' condition after endovascular occlusion of bronchial arteries improved so much, that this conservative treatment became sufficient, without the need of surgical intervention.

The reason for fading of the inflammatory process was the decrease of exudation degree, leucocyte release and reduction of secretion from bronchial glands. It should be emphasized that in our observations after bronchial artery embolization, we did not see destructive and necrotic changes, either in the bronchial wall, or in pulmonary parenchyma. It is connected with the presence of anastosomes at the pre-capillary level between the system of bronchial arteries and the pulmonary arterial system. The increase of the bronchial wall sclerosis and pneumosclerosis in the zone of pathological focus was observed in the late time after embolization. It indicated the completion of the cycle of pathological process. Thus, clinical observations give evidence to the fact, that hydrogel emboli can be considered as a material, which is absolutely suitable and can be successfully used for therapeutic occlusion of blood vessels [5-8].

References

[1] M.I. Kusin, Yu.D. Volynski, *Surgery*, 1978, 131-136.

[2] M.I. Kusin, Yu.D. Volynski, A.A. Vishnevskiy, P.I. Todua, *The Endovagcular Therapy and Surgery of Pulmonary Hemoptysis*, Irkutsk, 1981, (in Russian)

[3] Remy, C. Voisin, C. Dupruis, *Am. Radiology*, **17,** 1976, 5-16.

[4] N.E. Scuba, F.I. Todua, *The Endovascular Therapy and Surgery of Pulmonary Hemoptysis*, Irkutsk, 1981, (in Russian)

[5] Yu.V. Moiseev, G.E. Zaikov, *Chemical resistance of polymers in reactive media,* Plenum Press, New York, 1987, 586 pp.

[6] G.E. Zaikov, A.L. Iordanskii, V.S. Markin, *Diffusion of electrolytes in polymers,* VNU Science Press, Utrecht, 1988, 328 pp.

[7] G.E. Zaikov, *Degradation and stabilisation of polymers,* Nova Science Publ., New York, 1999, 198 pp.

[8] Jimenez, G.E. Zaikov, *Polymer analysis and degradation,* Nova Science Publ., New York, 2000, 287 pp.

5. Antitumor Agent Methotrexate-Containing p(HEMA)

5.1 Introduction

The widespread occurrence of malignant tumors and often low response rate of current methods of treatment call for new therapeutic approaches. Testing the effectiveness of a targeted introduction of cytostatic drugs into the neoplastic tissue has been reported in the literature [1, 2]. For this purpose endolymphatic introduction of cytostatics was performed or they were infused into the artery which supplies blood to the tumor tissue [3]. However, these procedures for targeted introduction of cytostatics into tumor tissues bring only a short-term effect of the drug unless the catheter is left in the vessel for a long time. A pronounced therapeutic effect usually requires that the procedure must be repeated many times. Such treatment with cytostatics is accompanied by the same toxic side effect on haemopoiesis, on the gastrointestinal tract and other organs as conventional treatments. Finally, since most of the drug given to a patient is "wasted", unnecessarily high amounts of expensive drugs must be administered.

An attractive procedure is that in which the cytostatic drug does not pass through blood or lymphatic vessels of the tumor but its effective concentration is maintained in them for a prolonged time using a single dosage form [4]. This may be achieved by a continuous controlled release, at biological pH, of the reversibly immobilized antitumor agent from a matrix. This can strongly increase the efficiency of the drug while limiting the toxic effects [4]. Among immobilization techniques, reversible electrostatical banding of the drug to a hydrophilic polymer matrix is recommended [5]. In addition to the effect of cytostatic drugs, slowing down of the growth of the tumor may be helped by decreasing blood supply by means of embolization of the supply vessels. A combination of these two complementary procedures seems justified. Moreover, in the preoperative embolization of blood vessels supplying the tumor these joint effects might help to decrease the release of tumor cells into the blood stream during the following operation and reduce their viability.

To put this into effect, we load emboli intended for occlusion of the tumor-supplying vessels with a cytostatic drug. The drug situated in the immediate vicinity of the tumor tissues should slowly diffuse into the tumor and probably ensure a prolonged suppression of its growth. Considerable experience exists in the application of emboli made from the poly(2-hydroxyethyl methacrylate) (p(HEMA)) hydrogel for the endovascular occlusion of blood vessels in the treatment of various diseases [6, 7]. Emboli made from the hydrogel are optimal from the therapeutic viewpoint since they prevent recanalization [8], important in ensuring long-term ischaemia of the tumor. Additionally, experience has been gathered with p(HEMA)-based spherical emboli as drug carriers, especially for haemostatic drugs (thrombin [9], ethamsylate [10], aminocaproic acid [10]).

In this Section we examine the possibility of loading emboli made from p(HEMA) hydrogel with one of the widely applied cytostatics - methotrexate, N-(4-(N-methyl-2,4-diamino-6-pteridinyl-methylamino)benzoyl) glutamic acid, the folic acid antagonist. This compound has been in clinical use for the treatment of various malignant as well as non-neoplastic diseases for many years. Furthermore, we investigate the character of histological changes in the vessel wall and in the surrounding tissues following the endovascular implantation of such biologically active emboli.

5.2 Materials and Methods

The synthesis of spherical p(HEMA) particles and the method of their purification have been described [8]. The beads used in this investigation were 0.4 - 0.6 mm in size in the dry state. A fraction (1 g) of these particles of medical grade purity was preswelled in dioxane, immersed in a solution of 0.358 g of 4-nitrophenyl chloroformate in 10 ml of dioxane, and the mixture was shaken at room temperature for 4.5 h. The unreacted reactants were removed by thorough washing with dioxane and the activated polymer was added to a solution of 1.5 g of 1,6-diaminohexane in 5 ml of dioxane with shaking at room temperature for 2.5 h. After completion of the reaction the p(HEMA)-Hex beads were thoroughly washed with dioxane, with 0.5 M sodium carbonate until the polymer lost its yellow color, and with water. The polymer particles, which according to elemental analysis contained 0.75 % (w:w) of nitrogen, were then immersed in 50 ml of 1.8 % (w:w) aqueous solution of methotrexate (MTX), which was dissolved by adding sodium hydrogen carbonate.

The desorption kinetics of MTX was determined spectrophotometrically (at 234 nm) in a 10 ml cell; water was changed after incubation of 5, 15, 25, 40 and 60 min. After this time there is virtually no further release of MTX. The amount of antimetabollite ionically bound on 1 g of the carrier was 0.19 g and was determined from the balance between the initially loaded and released amounts of antimetabollite. Sterilization of the samples was carried out in boiling distilled water for 45 min.

Donor blood analysis was based on measurements of changes in the number of leucocytes, erythrocytes and thrombocytes in 1 μl of blood, followed by determination of the haemolysis level increase [8].

A second method of the *in vitro* determination of haemostatic properties of materials consisted of an evaluation of indicators of donor plasma clotting in contact with the material: activated partial thromboplastin time (APTT), thrombin time (TT) and prothrombin time (PT). All these indicators were determined using a Hyland coagulometer in the donor plasma, to which 0.1 ml of a suspension of hydrogel particles in water (1:3) was added, i.e. in a ratio analogous to that introduced into the vessel through a catheter during the embolization.

The method of implantation of emboli into the rabbit's femoral artery has been described in a preceding paper [8].

A histological investigation of the blood vessels containing the emboli and of the surrounding tissues was carried out 1, 7 and 14 days after the implantation. Celloidine histological samples were stained with haematoxylin, eosin, and according to van Gieson.

5.3 Results and Discussion

5.3.1 Synthesis of p(HEMA)-Hex-MTX Particles

To avoid the decrease of methotrexate activity by its chemical modification we used simple sorption as a method of fixation to the polymer. In this report p(HEMA), the copolymer of 2-hydroxyethyl methacrylate (98 %, vol.) and ethylene dimethacrylate (2 % vol.) in the form of regular spherical porous particles was used as a polymeric carrier. A p(HEMA) carrier as such does not possess the required affinity for MTX or for other cytostatic substances [11]. From MTX solution-saturated p(HEMA), the drug can be quickly and completely eluted with water. It is known that the desired sorbing ability can be attained by introduction of aminohexyl groups to p(HEMA) [11]. Such a p(HEMA)-Hex derivative has been prepared from p(HEMA) by esterification of the hydroxyl groups with nitric acid, which is a dangerous and difficult reaction, followed by a reaction with 1,6-diaminohexane [12]. Other methods for the synthesis of amino derivatives of p(HEMA) are less direct. These include the reaction of an alkylating derivative of p(HEMA), e.g., halide, tosylate, or sulphate, with ammonia or an amine, or a reaction of p(HEMA) with alkylating amines, or activation with cyanogen bromide followed by substitution of active groups with airline. These methods employ compounds which are unsuitable for medical applications because of toxicity, instability, difficult manipulation, etc. To avoid these shortcomings, it is advantageous to use solid activated esters of chloroformic acid [13, 14].

In this Section, hydroxy groups of poly(2-hydroxyethyl methacrylate) (1) were activated [13, 14] in a nonaqueous solvent using 4-nitrophenyl chloroformate (2) according to:

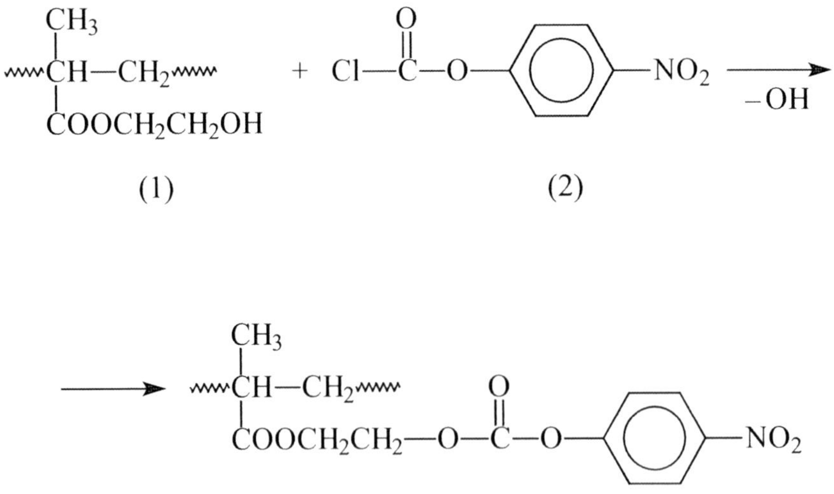

(1)

(3)

The resulting unstable product (3) was reacted with an amino group of 1,6-diaminohexane (4) to form the urethane bond according to:

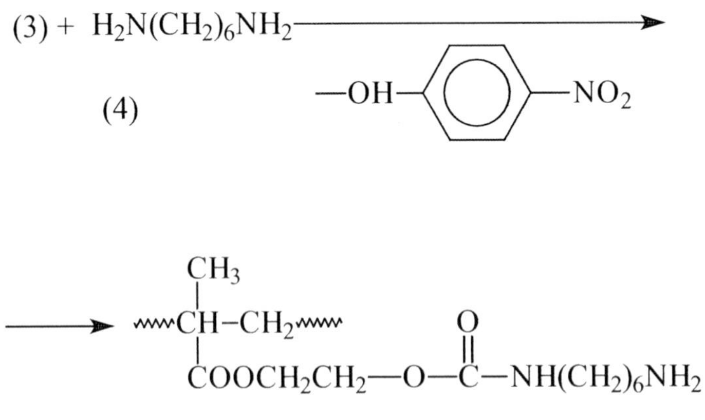

(2)

(5)

This activation of hydroxy groups of p(HEMA) leads to products which are easy to prepare, stable and non-hazardous. Another advantage is the easy check of binding, deactivation, and washing due to the yellow color of 4-nitrophenol. The aminohexyl derivative of poly(2-hydroxyethyl methacrylate) (p(HEMA)-Hex) (5) thus formed is a weakly basic anion exchanger, characterized by the total exchange capacity, capable of adsorbing MTX from aqueous solution. An alternate mechanism of MTX-fixation, e.g. reversible adsorption, cannot be excluded. At pH 6.8 the quantity of sorbed MTX was linearly proportional to its concentration in the aqueous solution up to a concentration given by the solubility limit of MTX in water at pH 6.8. The amount of MTX absorbed from such a solution was 0.21 g/g of dry polymer. The release of MTX in the organism might be based on the anion exchange. The biological environment is rich in various anions which can replace MTX from the p(HEMA)-Hex-MTX ion exchanger.

5.3.2 Haematolagical Testing

p(HEMA)-Hex-MTX spherical particles intended for embolization were kept in an excess of saturated MTX solution and then washed to medical grade purity before being used in experiments [8]. Donor blood analyses were employed as the criteria in evaluating leaching of excessive MTX from the emboli. In this test the morphology of the red blood cells and of elements of white blood (thrombocytes, lymphocytes, erythrocytes) was carefully examined. Blood was examined before and after contact with the hydrogel samples tested.

The data thus obtained showed that the first test set of unwashed p(HEMA)-Hex-MTX emboli causes rough changes in blood (haemolysis of blood elements, leucocytopenia, anemia, thrombocytopenia and a morphological damage to blood cells), which indicate tissue incompatibility. Redundant (unbound) MTX must be desorbed to obtain medical grade particles, because it could produce vast necroses of the artery wall after its embolization. In the p(HEMA)-Hex-MTX hydrogel there are apparently two kinds of sorbed MTX molecules: those that are immobilized by ionic bonds and those that are not. Such immobilization may involve a bond between acid carboxylic substituents of the MTX molecule with basic amino groups of the p(HEMA)-Hex polymer.

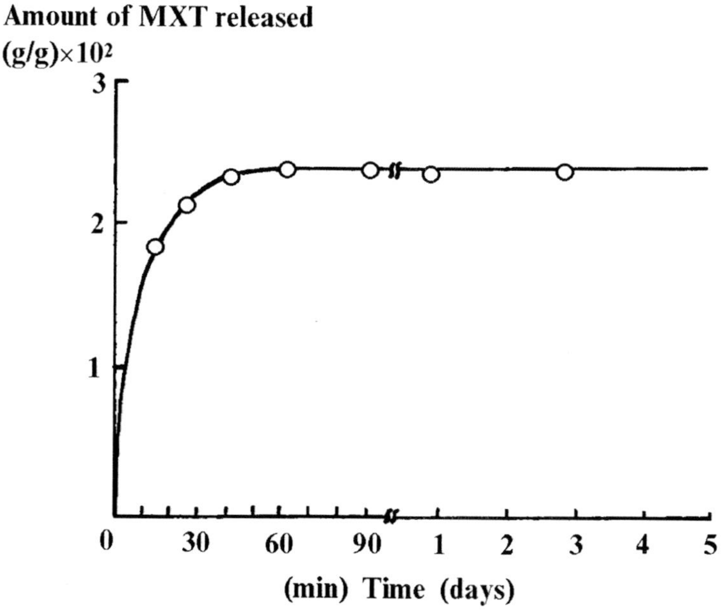

Figure 5.1. Desorption of MTX from PHEMA-Hex-MTX carrier as a function of time.

To remove redundant MTX, p(HEMA)-Hex-MTX spherical particles were placed in distilled water at room temperature, and the release rate was checked using the kinetics of desorption (Figure 5.1).The curve is characterized by a large initial slope corresponding to the fast release rate which takes 40 min and marks the elimination of MTX molecules in excess of the total capacity of the p(HEMA)-Hex polymer. This is followed by a stationary state which presumably signals completion of the elimination of MTX molecules exceeding the exchange capacity. The amount of such MTX molecules released over the whole time per gram of sample and determined by UV spectroscopy was quite small: 2×10^{-2} g/g. Here, it is worth mentioning that if p(HEMA) emboli are treated with a solution of ethamsylate, which is not very firmly bound with the hydrogel, the drug is desorbed from the hydrogel at a concentration higher than that in the case of MTX by one and a half orders of magnitude [10]. Calculations carried out using the kinetic curve data (Fig. 5.1) show that after purification by washing with distilled water the p(HEMA)-Hex-MTX polymer contains 0.19 g of ionically bound MTX per 1 g of the polymer. The efficiency of MTX binding with the p(HEMA)-Hex hydrogel from the MTX saturated solution is 24%). Haemolysis tests after contact of the purified p(HEMA)-Hex-MTX material with donor blood show minimal damage to the blood cells and an increase in thrombocytopenia at the expense of the formation of thrombocytic aggregates. Such material satisfies toxicological requirements and can be used in the endovascular occlusion.

The method described above, i.e. an analysis of changes in the donor blood elements in contact with the material, is to be used as a technological check of the degree of purification of biologically active emboli after MTX loading for their experimental and, in the future, clinical application.

5.3 Haemostatic Properties

The effect on blood coagulation by purified p(HEMA)-Hex-MTX emboli was investigated by means of a coagulogram, which is a general test used for determination of the effect of drugs on the blood coagulation system. Table 5.1 shows some indicators of the coagulogram: APTT, TT and PT both in healthy and in pathologically altered donor blood, and in the presence of MTX-containing and MTX-free hydrogel emboli. The MTX-free emboli were studied since the hydrogel has a haemostatic effect due to the nature of the monomer and the structure of p(HEMA) [8]. Pathological plasma was taken from patients suffering from a focal alteration of the liver, where APTT is extended from 69 to 90 s (at the APTT standard 30 - 35 s).

As documented by Table 5.1, the introduction of MTX does not affect APTT, if the plasma is healthy. In the case of pathologically altered plasma characterized by hypocoagulation there is a considerable (virtually double) decrease in APTT compared to the standard value. A still more positive effect following the introduction of MTX can be seen in a three-fold decrease in the TT indicators (patient No. 5 and 6). Results obtained from the interaction between the p(HEMA)-Hex-MTX hydrogel and blood show that the former exerts a weaker haemostatic effect compared with the hydrogel treated with ethamsylate or aminocaproic acid [10].

Table 5.1. Indicators of Donor Plasma Used in the Characterization of Properties of the PHEMA-Hex-MTX Hydrogel.

Indicator	Reported data for normal blood	Experimental data				
			No. of donor	Standard	PHEMA	PHEMA-Hex-MTX
APTT(s)	30 - 35	Healthy plasma	1	30	28	28
			2	31	28	32
			3	32	30	31
			4	29	25	23
		Pathological plasma	5	82	71	34
			6	89	76	58
			7	90	70	54
			8	69	58	27
TT(s)	16 - 21	Healthy plasma	1	24	18	17
			2	26	25	25
			3	22	18	18
			4	17	14	13
		Pathological plasma	5	105	100	32
			6	110	106	33
			7	112	82	29
			8	89	75	55
PT(s)	12	Healthy plasma	1	15	15	18
			2	18	18	17
		Pathological Plasma	5	26	—	18
			6	30	28	27
			8	32	27	30

APTT is the activated partial thromboplastin time; TT the thrombin time; PT the prothrombin time.

5.4 Histomorphological Investigation

The most important medico-biological test consisted of a histological investigation of p(HEMA)-Hex-MTX embolic material implanted in femoral arteries of a rabbit for various periods of time. Samples of vessels with surrounding tissues containing hydrogel emboli of the same size, but MTX-free, taken at the same times were used as control. It was shown that in control experiments the reaction of the vessel wall was virtually absent; only extension of the wall occurred if the latter was tightly filled with the emboli.

The implantation of p(HEMA)-Hex-MTX emboli caused necrosis of the intimate and a complete or partial necrosis of the muscular envelope of the artery in the section where the embolus adhered to the wall of the vessel. As a result, the emboli had adhered directly to the adventitia (Figs. 5.2(a-c)). In the skeletal muscles surrounding the vessel moderate necrosis was also observed within the times of investigation (Fig. 5.3a), this was followed by regeneration (Fig. 5.3b). With increasing time, absorption of necrotized muscular fibrils takes place, and their regeneration phenomena increase. Lysis and regeneration of the muscular tissue took place first near the blood vessel with MTX-containing emboli. Starting from the 7th day after the implantation, one part of the emboli appeared microporous without any cells in the pores of the material (Fig. 5.2b). On the other hand, in MTX-free emboli fibroblasts which had penetrated into the pores were found in their lumina; the number of the fibroblasts increased with increasing time of observation.

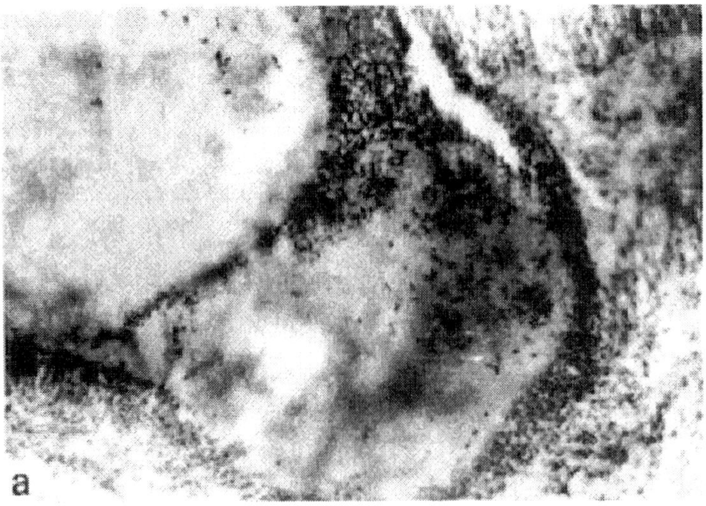

a - 1 day after the implantation. Artery wall in a state of necrobiosis and necrosis with massive infiltration of leucocytes. Between emboli detritus from necrotized intima.

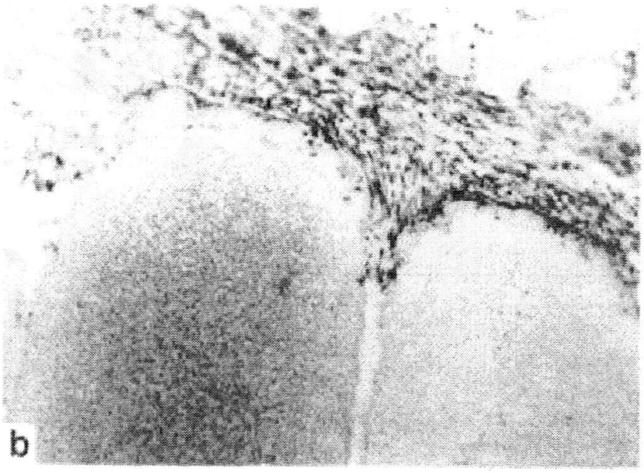

b - 7 days after the implantation. Intima and muscular envelope of the artery are lacking. The embolus adheres to the adventitia of the vessel. A small focus of the muscular envelope remains in the angle between two emboli. No penetration of cells into pores of the emboli.

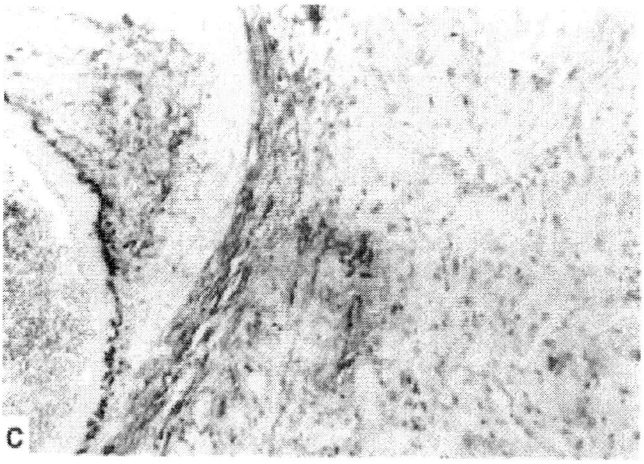

c - 14 days after the implantation. Intima and media necrotized. Sclerosis of the tissue in the perivascular space.

Figure 5.2. Histological section of the rabbit's femoral artery with two emboli (one embolus) from the p(HEMA)-Hex-MTX hydrogel. Stained after van Gieson: a, with haematoxylin and eosin, b, c. (× 160).

The results obtained suggest that in the animal MTX does not remain in an immediate contact with the material of the embolus. It is gradually washed out by the blood plasma and diffuses into the vessel wall and the surrounding tissues. The MTX activity is maintained, as indicated by necrotic changes in the vessel wall and surrounding tissues. After some time the necrotic parts are resorbed throughout the wall and in skeletal muscles. At the same time, the acute inflammation in the vessel wall recedes and changes to chronic inflammation in the tissues surrounding the vessel. It is noteworthy that fibroblasts grow into pores of the hydrogel, particularly by the 14th day of implantation. This indicates a sufficiently complete release of the cytostatic from the embolic material by this time and the absence of a local toxic effect, which allows cells of the connective tissue to grow unhampered into the pores of the hydrogel.

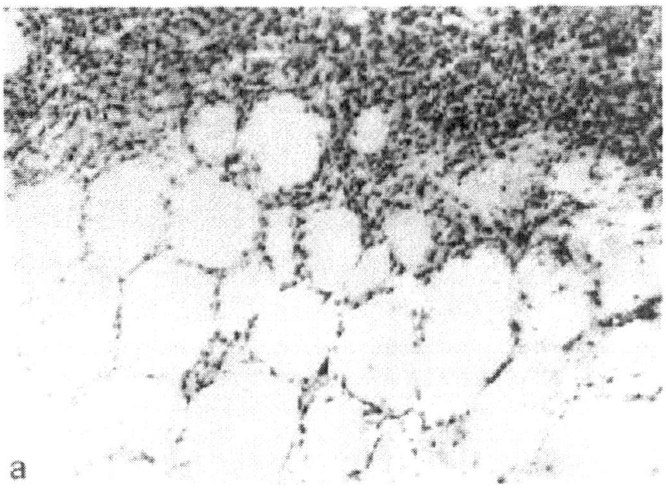

a - 7 days after the implantation. Necrosis with demarcation of inflammation.

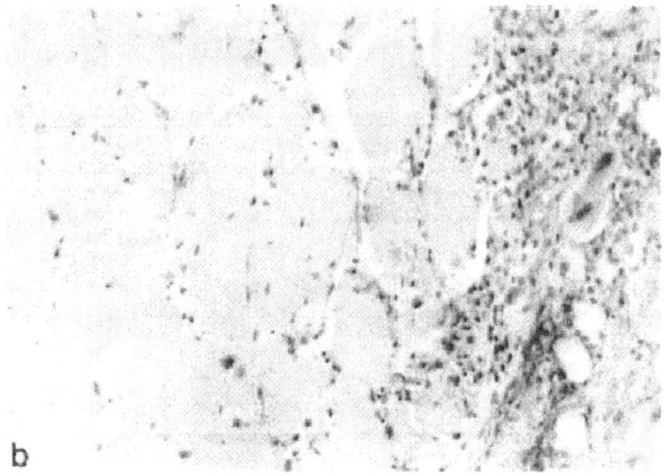

b - 14 days after the implantation. Necrosis and onset of regeneration.

Figure 5.3. Skeletal muscular fibers in the immediate vicinity of the p(HEMA)-Hex-MTX emboli. Stained with haematoxylin and eosin, a, after van Gieson, b. (× 160).

Preliminary data reported in this Section allow us to assume that emboli made from the modified p(HEMA) hydrogel can be loaded with methotrexate, one of the most frequently used cytostatics. Such a simple MTX-fixation to a p(HEMA)-Hex carrier has the advantage that the pharmacological activity of the drug remains unchanged. The use of p(HEMA)-Hex-

MTX hydrogel emboli should enable us to achieve a maximum concentration and a prolonged effect of the cytostatic in the embolized region (tumor) while reducing its concentration in other tissues of the organism at the same time. Thus, there is a prospect of the targeted local palliative cytostatic effect of the drug on the tumor and of prevention of the possible haematogenic formation of metastases, along with a decreased blood supply to this zone or to the whole organ. In other words, site-specific function of released antitumor drug could be combined with the endovascular occlusion of blood supply to the tumor tissue [15, 16].

References

[1] *Intra-Arterial and Intfacavitary Cancer Chemotherapy*, Ed. S.B. Howel, Martinus Nijhoff Publishers, Boston, 1984.

[2] *Regional Cancer Treatment*, Eds. K.R. Aigner, Y.Z. Patt, K.H. Link, J. Kreidler, Karger, Munchen, 1988.

[3] M.C. Sheen, Intra-arterial Infusion Chemotherapy for Advanced Head and Neck Cancer, in *Advances in Regional Cancer Therapy*, Eds. J. Kreidler, K.H. Link, K.R. Aigner, Karger, Munchen, 1988, 144-150.

[4] J. Kost, R. Langer, Equilibrium Swollen Hydrogels, in *Controlled Release Applications, in Hydrogels in Medicine and Pharmacy*, Ed. N.A. Poppas, CRC Press, Boca Raton, 1987, **3**, 95-108.

[5] W.R. Gombotz, A.S. Hoffman, Immobilization of Biomolecules and Cells on and within Synthetic Polymer Hydrogels, in *Hydrogels in Medicine and Pharmacy*, Ed. N.A. Poppas, CRC Press, Boca Raton, 1986, **1**, 95-126.

[6] D. Horak, F. Svec, J. Kalal, A. Adamyan, Yu. Volynskii, O. Voronkova, L. Kokov, K. Gumargalieva, Hydrogels in Endovascular Embolization. II. Clinical Use of Spherical Particles, *Biomaterials*, 1986, **7**, 467-470.

[7] D. Horak, F. Svec, A. Adamyan, M. Titova, O. Voronkova, N. Trostenyuk, N., Vishnevskii, V., Guseinov, E. and Gumargalieva, K., Poly[2-hydroxyethyl methacrylate) beads for the preoperative endovascular occlusion of branches of the hepatic artery in focal alterations of the liver, *Call. Mater*. 1990, **6**, 287-297.

[8] D. Horak, F. Svec, J. Kalal, K. Gumargalieva, A. Adamyan, N. Skuba, M. Titova, N. Trostenyuk, Hydrogels in Endovascular Embolization. I. Spherical Particles of Poly(2-hydroxyethyl methacrylate) and Their Medico-biological Properties, *Biomaterials*, 1986, **7**, 188-192.

[9] D. Horak, F. Svec, J. Kalal, A. Adamyan, M. Titova, N. Trostenyuk, N. Skuba, V. Dan, O. Voronkova, K. Gumargalieva, Biologically Active Thrombin-containing Hydrogels Based on Poly(2-hydroxyethyl methacrylate) for Endovascular Occlusion, *Polymers in Medicine*, 1991, **21**, 31-41.

[10] D. Horak, F. Svec, A. Adamyan, M. Titova, N. Skuba, O. Voronkova, N. Trostenyuk, V. Vishnevskli, K. Cumargalieva, Haemostatic Activity of Ethamsylate and Aminocaproic Acid Adsorbed Poly(2-hydroxyethyl methacrylate) Particles, *Biomaterials*. (in press)

[11] K. Motycka, K. Slavik, A. Kocovska, R. Cihar, P. Spacek, M. Kubin, Effect of Methotrexate Sorbed on Modified 2-Hydroxyethylmethacrylate Carriers in Mice of C3H Strain with a Solid Gardner Lymphosarcoma, *Neoplasma*, 1977, **24**, 271-276.

[12] P. Spacek, M. Kubin, M. Bones, *The Method of Production of Polymeric Esters of Methacrylic Acid and Multifunctional Alcohols*, Czech. Pat. No. 191474, 1975.

[13] J. Drobnik, J. Labsky, H. Kudlvasrova, V. Saudek, F. Svec, The Activation of Hydroxy Groups of Carriers with 4-Nitrophenyl and N-Hydroxysuccinimidyl Chloroformates, *Biotechnol. Bioeng.*, 1982, **24**, 487-493.

[14] J. Drobnik, J. Kalal, J. Labsky, V. Saudek, F. Svec, *The Method of Activation of Insoluble Carriers Containing Hydroxy Groups*, Czech. Pat. No. 204190, 1980.

[15] A.Popov, N. Rapoport, G. Zaikov, *Oxidation of stressed polymers,* Gordon and Breach, New York, 1991, 336 pp.

[16] A.F. Bochkov, G.E. Zaikov, V.A. Afanasiev, *Carbohydrates,* VSP Science Press VB, Zeist-Utrecht, 1991, 154 pp.

6. Poly(2-Hydroxyethyl Methacrylate) with Intensified Haemostatic Activity as a New Embolic Material

Embolization of arterial vessels is increasingly used in surgery and clinical oncology [1]. It may be indicated by a threatening hemorrhage from a decomposing tumor, the necessity of creation of a "dry" operative field or the need to reduce the release of tumor cells into the bloodstream during surgical intervention. In the case of nonresectable tumors, embolization of the supply vessels often slows the growth of the neoplasm, postpones formation of metastases, removes painful paraneoplastic syndromes and provides favorable conditions for the subsequent radiation or drug treatment with palliative aims.

Earlier, we demonstrated the prospects and justification of endovascular embolization in haemostasis and as a prophylactic measure for bleeding, using emboli made from poly(2-hydroxyethyl methacrylate) (PHEMA) hydrogel [2, 3, 6], which is capable of primary activation of thrombocytic haemostasis, reduction in activity of fibrinolysis and secondary activation of the plasma haemostasis. The thrombogenic properties of PHEMA hydrogels were affected by three factors [4]: degradation products of the hydrogel, adsorption of plasma proteins by the hydrogel, and the structure of the implant surface. Endovascular embolization by means of PHEMA is justified as a preparatory step before surgical interventions in zones of increased vascularization in cases of hypocoagulation, where no immediate haemostasis is required.

To achieve an immediate and efficient management of bleeding from a variety of origins, further development of new composition of the emboli, having increased haemostatic properties, is needed. This may be accomplished by immobilization or sorption of biologically active compounds.

In an earlier study we reported the preparation and properties of thrombin-containing PHEMA hydrogel [5]. The immobilization or sorption of a coagulating enzyme (thrombin) to the PHEMA emboli strengthened their haemostatic properties and activated the coagulation system of the organism. However, the thrombin used in the preceding study showed several draw backs: it may stimulate allergic reactions and, if applied in the wrong way, the activation of the whole coagulation system can be pathological. This is probably due to the fact that, because of the high costs of purified thrombin, an unpurified enzyme was used. For this reason, haemostatics, such as ethamsylate and aminocaproic acid, whose effect is not identical with that of thrombin, were used in this study to intensify the haemostatic properties of PHEMA emboli. Ethalnsylate (Dicynone in Switzerland and other countries) is a well-known synthetic water-soluble non-steroidal haemostatic agent (diethylammonium 1,4-dihydroxy-3-benzenesulfonate) with a broad range of activity. Unlike thrombin, ethamsylate can be used in the treatment of capillary bleedings without stimulation of allergic reactions. It exerts its action on primary haemostasis and is effective in restoring platelet functions, e.g. reduces bleeding time in patients with thrombopathies (platelet dysfunctions). Essentially, the effect of ethamsylate consists in that it helps: a) raise the number of thrombocytes, b) increase their adhesiveness, and c) increase capillary strength by reinforcement of the capillary basal membrane while shortening the bleeding time.

Aminocaproic acid (6-aminobexanoic acid), as a synthetic inhibitor of proteolysis, stimulates the blockage of proteolytic enzymes while simultaneously inhibiting the effect of plasmin (fibrinolysin) on fibrinogen and coagulation factors. By blocking activators of plasminogen and partly suppressing the effect of plasmin, aminocaproic acid has a specific haemostatic effect in hemorrhages connected with increased fibrinolysis. Unlike ethamsylate, which in tensifies the aggregation of thrombocytes, aminocaproic acid has no effect on their aggregation.

6.2 Materials and Methods

6.2.1 Synthesis

The synthesis of spherical PHEMA particles and the method of their purification have been described [2]. The beads used in this investigation were 0.4 - 0.6 mm in size in the dry state. A part (1 g) of these particles of medical grade purity was immersed in a 16 ml of a 0.125 wt.% aqueous solution of ethamsylate (Dicynone, Lek, Ljubljana, Yugoslavia) for 6 days. The amount of ethamsylate sorbed on 1 g of carrier amounted to 0.87 g which represents 43.5% of the ethamsylate in the initial solution; this was determined from the weight balances prior and after ethamsylate sorption.

Another part, (1 g) of above mentioned PHEMA particles was immersed in a 16 ml of a 5% aqueous solution of aminocaproic acid (Acikaprin, Polfa, Warsaw, Poland) and left there for three hours.

6.2.1 Purification Methods

In a 10 ml cell was placed 0.1 g of the spherical emboli which had been treated with ethamsylate. Water was changed after incubation times of 5, 15, 25, 40, 60 min. Washings took place under spectrophotometric control until the re lease of the drug had ceased. The spectroscopic measurements were carried out at 215 nm with a Pye-Unicam SP 100 apparatus. The kinetic curve of drug release from the spherical particles is obtained as a result of summation of the amount of drug released within each time interval. The amount of ethamsylate sorbed on 1 g of the hydrogel was determined from the weight balance of ethamsylate in solutions prior and after washing.

6.2.2 Biocompatibility Tests

Analysis of the donor blood before and after contact with the polymeric materials in terms of number of leucocytes, erythrocytes and thrombocytes was used as a criterion of the quality of emboli purification. The morphology of thrombocytes and erythrocytes was also examined. Haemostatic properties of the emboli were characterized in terms of donor plasma clotting: the activated partial thromboplastin time (APTT), the thrombin time (TT) and the prothrombm time (FT). All of these indicators were determined using a Hyland coagulometer in the donor plasma, to which 0.1 ml of a suspension of hydrogel emboli in water (1 : 3) was added. This is a ratio analogous to that introduced into the vessels during the embolization. The control consisted in the addition of water (without particles) which diluted the plasma and thus prolonged both APTT and TT.

6.3 Results and Discussion

In an attempt to prepare an embolic material suitable for efficient management of massive bleeding, spherical PHEMA particles used in earlier endovascular occlusion studies [3] were used as the matrix for sorbing haemostatics (ethamsylate or aminocaproic acid).

For any modification of PHEMA emboli, preservation of their biological compatibility must be examined. Investigation of the interaction between native blood and unwashed spherical hydrogel immediately after treatment with ethamsylate solution showed that this embolic material causes a substantial pathological change in blood (acute haemolysis, leucocytopenia, thrombocytopenia, anemia and morphological damage to blood cells). This would indicate potential tissue incompatibility. Indeed, histomorphological investigation of excess of ethamsylate-containing emboli implanted in the rabbit's femoral artery revealed necrosis of the infcima and of the muscular skeletal envelope of the vessel surrounding the hydrogen.

For the purpose of subsequent clinical application a washing of the emboli to achieve desorption of the excess of ethamsylate is required if medical grade particles are to be obtained. To remove excess ethamsylate, the ethamsylate-sorbed emboli were placed in distilled water at room temperature; the release rate was obtained by using the kinetics of desorption (Figure 6.1). The curve is characterized by a large initial slope corresponding to a fast release rate (50% of unbound ethamsylate is released within 10 min) and is followed by a stationary state, which should indicate the complete elimination of redundant ethamsylate. Redundant ethamsylate-release is virtually completed within 40 min. This is achieved by the four to five changes of water at 10 min intervals at room temperature. Desorption of the excess of ethamsylate shows that this ethamsylate is sorbed in a less stable way than is e.g. methotrexate, which is released from the hydrogel at a concentration lower by one and a half orders of magnitude [7]. That is, in contrast to methotrexate, which is ionically bound to the aminohexyl derivative of PHEMA, ethamsylate appears to be attached to PHEMA via hydrophobic interactions. Calculations carried out using the kinetic curve data (Figure 6.1) showed that after purification by washing with distilled water the hydrogel contained 0.33 g of sorbed ethamsylate per g of the polymer. The efficiency of ethamsylate binding with the PHEMA hydrogel from the ethamsylate saturated solution is 16.5%.

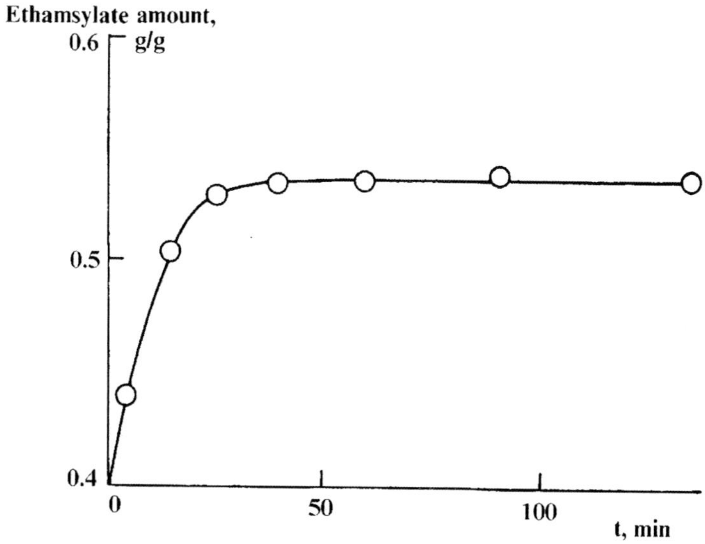

Figure 6.1. Desorption of redundant ethamsylate from spherical PHEMA particles as a function of time.

After purification of the polymer by washing with distilled water, the haemolysis of blood cells was retested and minimal damage was observed. All blood cells retained their natural character and no haemolysis took place. Thrombocytopenia was still observed with the washed ethamsylate-containing hydrogel; this was accompanied by the formation of thrombocytic aggregates (8 - 10 aggregates per 1000 erythrocytes). The aggregate of thrombocytes was bulkier than that seen with the ethamsylate-free PHEMA hydrogel, this indicates that ethamsylate is slowly released from the washed embolic material inducing aggregation of the thrombocytes. However, such material satisfies the toxicological requirements and thus can be used in the endovascular occlusion. The spectrophotometrically determined ethamsylate concentration in an external aqueous medium which does not stimulate haemolysis was 5×10^{-7} g ethamsylate/ml water.

In order to determine the *in vitro* haemostatic properties of the purified ethamsylate-containing PHEMA in contact with donor plasma, both thrombocytic and plasma haemostasis were characterized by determining the activated partial thromboplastin time (APTT, standard 30 - 35 s), thrombin time (TT, standard 16 - 21 s) and prothrombin time (PT, standard 12 s). The investigation concerned both normal plasma, of healthy donors and pathologically altered plasma of patients suffering from focal alterations of the liver (haemangiomas). In such patients the acute hypocoagulation reaction of blood is typical. Basically, this blood can be divided into two groups, namely, with less pronounced pathological changes (APTT 40 - 48 s, the standard being 30 - 35 s) and more pronounced pathological changes (APTT 69 - 90 s). The control consisted in the determination of indicators of plasma clotting, with distilled water added instead of the suspension of particles. The indicators did not exceed the standard very much, i.e. water has no haemostatic effect at all (Table 6.1).

Table 6.1. Indicators of Normal and Abnormal Plasma in Contact with Ethamsylate or Aminocaproic Acid-Containing Hydrogels.

		No. of donor	Standard	Water	PHEMA	Ethamsylate containing PHEMA	Aminocaproic acid containing PHEMA
APTT (sec)	Healthy plasma	1	32	36	23	16	8
		2	32	31	25	21	16
	Less pronounced pathological changes	3	45	52	24	15	5
		4	48	50	26	18	7
		5	40	42	30	24	14
	More pronounced pathological changes	6	82	—	71	24	
		7	89	—	76	43	
		8	90	—	70	46	
		9	69	—	58	22	
TT (sec)	Healthy plasma	1	20	24	18	14	10
		2	20	—	14	10	—
	Less pronounced pathological changes	3	35	38	17	15	14
		4	31	—	24	21	—
	More pronounced pathological changes	6	105	—	100	30	
		7	110	—	106	35	
		8	112	—	82	20	
		9	89	—	75	43	
PT (sec)	Healthy plasma	1	12	13	9	7	7
		2	12	—	10	9	—
	Less pronounced pathological changes	3	21	21	9	8	7
		4	24	—	14	13	8
	More pronounced pathological changes	6	26	—	—	16	
		7	30	—	28	20	
		8	—	—	—	—	
		9	32	—	27	19	

APTT is the activated partial thromboplasmin time; TT the thrombin time; PT the prothrombin time.

Abbreviated title: PHEMA is an embolic material.

As can be seen, the values of indicators of coagulogram for a healthy plasma in contact with ethamsylate-free and ethamsylate-containmg hydrogel are close to each other. This suggests that in the case of normal healthy blood one could use the hydrogel without its modification with ethamsylate (or aminocaproic acid) if endovascular embolization is indicated. However, in the group of patients with more distinct blood changes the haemostatic properties of free hydrogel will probably be unsatisfactory. In the case of blood with more or less distinct pathological changes the ethamsylate-treated emboli possess stronger haemostatic properties than the control starting ethamsylate-free emboli. Such materials accelerate the coagulation (APTT and TT) of pathologically changed blood whereas the blood clotting time had been slowed down and is almost three times that of normal blood. It should be noted that, while the APTT and TT reduction is very pronounced, the PT shows rather minor changes. Such reduction in APTT and TT indicated that ethamsylate, which is a general haemostatic, is effective even after being sorbed into the PHEMA hydrogel. The haemostatic effect of PHEMA is thus intensified.

The data shows the morphology of blood in contact with the hydrogel treated with a 5% aqueous solution of aminocaproic acid for 3 h. Since this solution is the same concentration as that used clinically in the intravenous infusion (and does not provoke haemolysis), the hydrogel treated with aminocaproic acid was not washed to remove any redundant acid. While the morphology shown by the data confirms that no haemolysis takes place, there is indication of a mild aggregation of thrombocytes (less than with ethamsylate) which can be explained by the effect of aminocaproic acid on blood plasma factors. The aminocaproic acid-treated hydrogel possesses a pronounced haemostatic effect (Table 6.1), although the latter was examined using only blood plasma with less pronounced pathological changes. Unlike the ethamsylate-treated hydrogel, not only the effect of decreased APTT, but also the effects of the other two decreased indicators of a coagulogram, TT and FT, can be seen.

Thus, we recommend PHEMA as the carrier of two drugs: ethamsylate, which increases thrombocyte adhesiveness and capillary resistance while affecting the thrombocyte-vessel wall interaction, and aminocaproic acid which is a synthetic inhibitor of proteolysis and an antiallergic substance. The etham- sylate or aminocaproic acid-treated PHEMA hydrogel has a particularly strong effect on the pathologically abnormal blood (i.e., accelerates the clotting time) of patients suffering from focal alterations of the liver. These patients have a disturbed synthesis of proteins and blood coagulation factors. Thus, this biologically active hydrogel is recommended for the endovascular occlusion treatment of haemangiomas of the liver associated with acute hypocoagulation. The use of this emboli with intensified haemostatic properties will reduce intraoperational blood loss from the liver tissue in acute parenchymatic hemorrhages, not only due to the blockage of blood vessels with the emboli, but also because of the effect of sorbed aminocaproic acid. The rate of release of haemostatics from PHEMA is due to the porous structure of the hydrogel and to the gradual washing out of the haemostatics by tissue liquids. The haemostatics may be effective in restoring factors of the disturbed plasma and thrombocytic haemostasis, and can mediate activation of reparatory processes in the dysfunction of the third stage of coagulation (the stabilization of fibrin when fibrin S is transformed into fibrin I). This extends the application of endovascular occlusion in the urgent treatment of various forms of hypocoagulation bleedings (parenchymatous bleeding of the liver, pulmonary bleeding, bleeding from stomach ulcer, acute bleedings in childbirth, including haemangiomas of the small pelvis), as well as in the treatment of pronounced allergic reactions.

The haemostatic properties of PHEMA emboli can be improved by sorption of ethamsylate or aminocaproic acid. The ethamsylate-sorbed emboli cause haemolysis and morphological changes of blood cells, unless they are purified to remove the redundant solution of ethamsylate. The threshold ethamsylate concentration, below which no haemolysis takes place, is 5×10^{-7} g/ml.

An analysis of donor blood in contact with the material (change in the number of erythrocytes, leucocytes, thrombocytes and their morphological changes) as well as determination of indicators of coagulogram can be used for the evaluation of haemostatic properties of new materials intended for contact with blood. Haemostatic properties of ethamsylate-sorbed hydrogel particles tested on normal healthy plasma were not very distinct. Tests carried out with pathologically changed plasma of the patients suffering from focal alter ation of the liver showed a pronounced haemostatic effect of ethamsylate or aminocaproic acid-containing hydrogels.

A comparison of haemostatic properties of drug-free PHEMA emboli with those ofethamsylate or aminocaproic acid-treated PHEMA indicate their potential in the treatment to prevent bleeding, particularly for those ca-used by a disturbed haemostasis system [8-11].

References

[1] D.J. Allison, Therapeutic Embolization and Venous Sampling, in *Recent Advances in Surgery*, Ed. Selwyn Taylor, Churchill Livingstone, 1980, 27-64.

[2] D. Horak, F. Svec, J. Kalal, K. Gumargalieva, A. Adamyan, N. Skuba, M. Titová, N. Trostenyuk, Hydrogels in Endovascular Embolization. I. Spherical Particles of Poly(2-hydroxyethyl methacrylate) and Their Medico-biological Properties, *Biomaterials*, 1986, **7**, 188-192.

[3] D. Horak, F. Svec, J. Kalal, A. Adamyan, Yu. Volynskii, O. Voronkova, L. Kokov, K. Gumargalieva, Hydrogels in Endovascular Embolization. II. Clinical Use of Spherical Particles, *Biomaterials*, 1986, **7**, 467-470.

[4] D. Horak, F. Svec, J. Kalal, A. Adamyan, N. Skuba, M. Titova, V. Dan, B. Varava, N. Trostenyuk, O. Voronkova, K. Gumargalieva, V. Timochina, Hydrogels in Endovascular Embolization. IV. Effect of Radiopaque Spherical Particles on the Living Tissue, *Biomaterials*, 1988, **9**, 367-371.

[5] D. Horak, F. Svec, J. Kalal, A. Adamyan, M. Titova, N. Trostenyuk, N. Skuba, V. Dan, O. Voronkova, K. Gumargalieva, Biologically Active Thrombin-containing Hydrogels Based on Poly(2-hydroxyethyl methacrylate) for Endovascular Occlusion, *Polymers in Medicine*, 1991, **21**, 33-43.

[6] D. Horak, F. Svec, A. Adamyan, M. Titova, O. Voronkova, N. Trostenyuk, V. Vishnevskii, E. Guseinov, K. Gumargalieva, Poly(2-hydroxyethyl methacrylate) Beads for the Preoperative Endovascular Occlusion of Branches of the Hepatic Artery in Focal Alterations of the Liver, *Can. Mater.*, 1990, **6**, 287-297.

[7] D. Horak, F. Svec, A. Adamyan, M. Titova, N. Skuba, O. Voronkova, N. Trostenyuk, V. Vishnevskii, K. Gumargalieva, Hydrogels in Endovascular Embolization. V. Antitumor Agent Methotrexate-containing PHEMA, *Biomaterials*. (in press)

[8] A.L. Iordanskii, T.E. Rudakova, G.E. Zaikov, *Interaction of polymers with bioactive and corrosive media*, VSP Int. Sci. Publ., Utrecht, 1994, 298 pp.

[9] A.Ya. Polishchuk, G.E. Zaikov, *Multicomponent transport in polymer systems*, Gordon and Breach, New York, 1996, 231 pp.
[10] G.E. Zaikov, *Biomedical application of polymers*, Nova Science Publ., New York, 1998, 212 pp.
[11] E.A. Parfenov, G.E. Zaikov, *Biotic type antioxidants*, VSP Inter. Sci. Publ., Zeist (The Netherlands), 2000, 560 pp.

In: Chemical Reactions in Liguid and Solid Phase
Editors: G. E. Zaikov and A. Jiménez, pp. 61-68
ISBN 1-59033-854-5

Chapter 3

OBTAINING OF SUSPENSIONS AT CARRYING OUT OF LIQUID-PHASE FAST CHEMICAL REACTIONS IN TURBULENT MODE

Vadim P. Zakharov[1], Irik V. Sadykov[2], Karl S. Minsker[1], Gennadi E. Zaikov[3] and Yuri B. Monakov[2]

[1]Bashkir State University, 450074, Ufa, Frunze str., 32,
Chemical Faculty, E-mail: ZaharovVP@mail.ru
[2]Institute of Organic Chemistry the Ufa Center of Science
Russian Academy of Sciences,450054, Ufa, Pr. Oktyabrya, 71
[3]Institute of Biochemical Physics of Science. Russian
Academy of Sciences, 117334 Moscow, Kosygin str. 4

ABSTRACT

In tubular turbulent devices for account of hydrodynamics movement of reagents is possible the reception of suspensions at passing fast chemical reactions with rather small radius of particles and factor of polydispersity, that is achieved in the stirred tank reactor with a mechanical agitator at speed of agitating more than 2000 rpm.

Key word/Phrase index. tubular turbulent devices, suspensions, barium sulphate, fast chemical reaction, precipitation.

In chemical technology the processes of reception suspensions by a condensation way are widely widespread at passing chemical reaction with precipitation. In many cases there is a task of reception homogeneous high dispersity of suspensions, in particular, at formation of high-activity heterogeneous and microheterogeneous catalytic systems Ziegler-Natta for polymerisation of olefines and dienes, antiagglomerates for synthetic rubbers, pigments for paints etc. In conditions of industrial production for carrying out of fast chemical reactions the way of reception of suspensions is widespread at mixture of feeding homogeneous reagents in a turbulent mode sold by blades of impeller pumps. However, probability of failure of motion elements of pumps in this case is great, their clogging, and at use of aggressive environments is necessary to apply express anticorrosive coverings of working surfaces of the device. Whereas it is usual that chemical reactions underlying to obtain high dispersity of suspensions, are ion-exchange and, as a rule, proceed with high speeds, it is possible to assume, that in this case it is expedient to use a new type of devices of chemical technology - small-sized high capacity tubular turbulent devices of a jet type [1]. The specifity of the work with these devices is enough well investigated and now they will be widely used in industrial production [2] at carrying out of fast homogeneous chemical reactions. However, there are no works concerning opportunities of their use for carrying out of fast chemical reactions at a combination of homogeneous liquid-phase flows to formation of suspensions in the literature.

The purpose of the present work was the study of the opportunity to use tubular turbulent devices a divergent-convergent of type for obtain homogeneous of suspensions by a condensation way at carrying out fast chemical reaction and comparison of efficiency of their work with stirred tank reactors at mechanical agitating.

As model the fast chemical reaction was used the reaction of sulphate barium precipitation:

$$H_2SO_4 + BaCl_2 \rightarrow BaSO_4\downarrow + 2HCl$$

Three methods of synthesis were used: interaction of barium chloride with sulfuric acid which is taking place in the stirred tank reactors (a three-necked flask in volume 500 cm^3, filled on 70%) at a rotating agitator working by a principle of an impeller pump and creating intensive turbulent flows in an axial direction reaction volume (variant 1, fig. 1, a); a method similar to variant 1, but in bulk reaction mixtures have been established the mixing baffle (variant 2, fig. 1, a), that alongside with the increase in capacity spent on agitating, increase turbulent mixture of flows [3]; interaction of feeding reagents (the water solutions H_2SO_4 and $BaCl_2$) carried out immediately in the six-section tubular turbulent device of divergent-convergent type (divergent diameter 24 mm, convergent diameter 10 mm, length of section 48 mm) at peripheral speed of movement of flows reagents $V \approx 0.5$ m/s (fig. 1, b). With the purpose of increase of the speed nucleation/ rate growth of crystals ratio and, as a consequence, the obtaining of the highest dispersity of systems, we have used concentrated (0.52 mol/l) and diluted (0.03 mol/l) water solutions $BaCl_2$ and H_2SO_4, accordingly.

Dispersion analysis of obtained suspensions $BaSO_4$ carried out by a method of sedimentation in a gravitational field with use torsion of weights on a standard technique [4]. For reduction of the probability of interaction between particles at sedimentation we obtained diluted solutions of suspensions $BaSO_4$ (~ 0.35% of weights). At data processing of sedimentation analysis we have used graphic differentiation of the accumulation curve of a deposit (fig. 2). For this purpose the sedimentation dependences in weight of the dropped out deposit m on time τ were approximated sigmoidal by three-parametrical function of a kind:

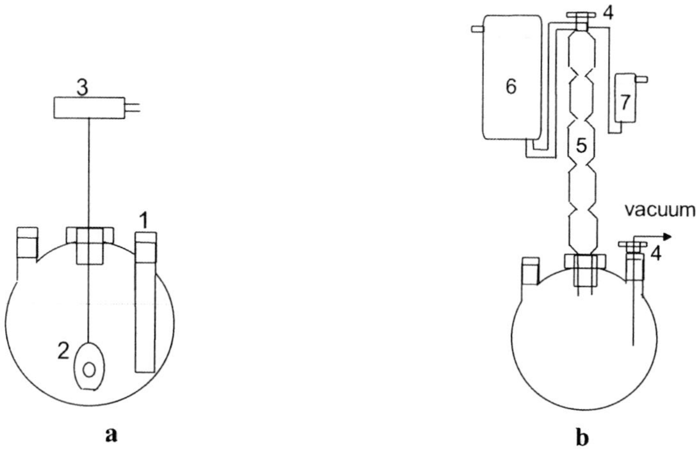

Fig. 1. Experimental installation. 1 - mixing baffle; 2 - agitator working by a principle of an impeller pump; 3 - electric drive of an agitator; 4 - three-way cock; 5 - tubular turbulent device; 6, 7 - container for solutions H_2SO_4 and $BaCl_2$, accordingly.

$$m(\tau) = \frac{a}{1 + \exp(-\frac{\tau - x_0}{b})} \quad (1)$$

The coefficients a, b and x_0 in (1), and also the dispersion at approximating the experimental sedimentation curves (fig. 2), are given in table 1.

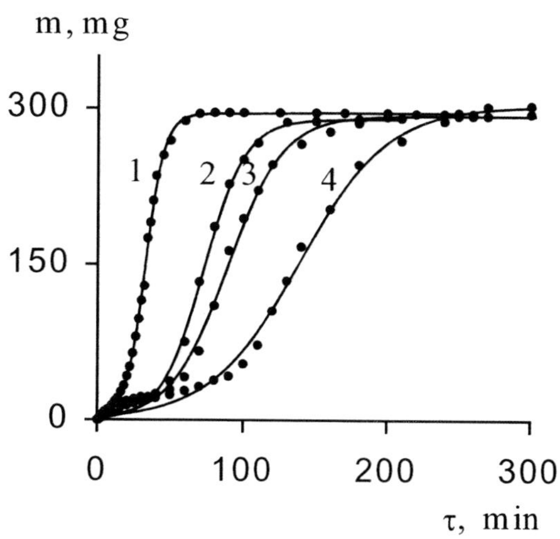

Fig. 2. Sedimentation curves of particles $BaSO_4$. Speed of agitation (variant 2): 300 (1); 1500 (2); 2500 (3) rpm; 4 - tubular turbulent device. Points - experiment; lines - approximating function.

Table 1. The Characteristics of the Approximation

Speed of agitating N. rpm	a	b	x_0	Coefficient of correlation R
	The stirred tank rector with mixing baffles (variant 2)			
300	295.1±2.1	6.6±0.2	32.4±0.2	0.99
1500	289.8±3.2	14.2±0.6	72.8±0.8	0.99
2500	293.3±1.8	19.1±0.6	88.9±0.8	0.99
	The tubular turbulent device			
	303.2±2.6	30.7±1.1	138.7±1.4	0.99

In this case, according to the equation Svedber-Oden [4], weights m_i of a fraction of particles suspension with radius $r>r_i$, laden down to the moment of time τ_i, paid off as:

$$m_i = \frac{a}{1+\exp(-\frac{\tau_i - x_0}{b})} - \tau_i \frac{\exp(-\frac{\tau_i - x_0}{b})a}{b(1+\exp(-\frac{\tau_i - x_0}{b}))^2} \tag{2}$$

The microscopic analysis of the obtained suspensions has shown that the $BaSO_4$ particles have brightly expressed anisometry (represent rods with the relation of length to a diameter of the order 4, that explains sigmoidal a kind of the sedimentation curves with an excess). In the initial moment of sedimentation the rotation "rods" of particles is possible, that determines padding resistance to its movement (similarly to increase of viscosity) and retardation of speed of accumulation deposit. Besides at free settling of a particle of the unspherical form are oriented in a direction of movement so that the maximal resistance of motion was created. It also reduces fall velocity of solids in a liquid and complicates definition of their true sizes. In this connection, by results of sedimentation analysis determined equivalent radius r_e (radius of a spherical particle settling with the same speed).

By the form of the sedimentation curves (fig. 2) and the differential curves (fig. 3) it is visible that an increase of speed of agitation and their displacement in area of particles of the smaller size (large times of sedimentation) is observed, i.e. an increase of the level of turbulent mixture in a zone of reaction there is a reduction of average radius of particles. Whereas the differential distribution of particles of dispersed phase on the sizes $F(r_e)$ (fig. 3) is the typical distribution function of casual size, for account most probable radius r_{mp} of particles suspension have used the following expression:

$$r_{mp} = \frac{\int_0^\infty r_e F(r_e) dr_e}{\int_0^\infty F(r_e) dr_e} \tag{3}$$

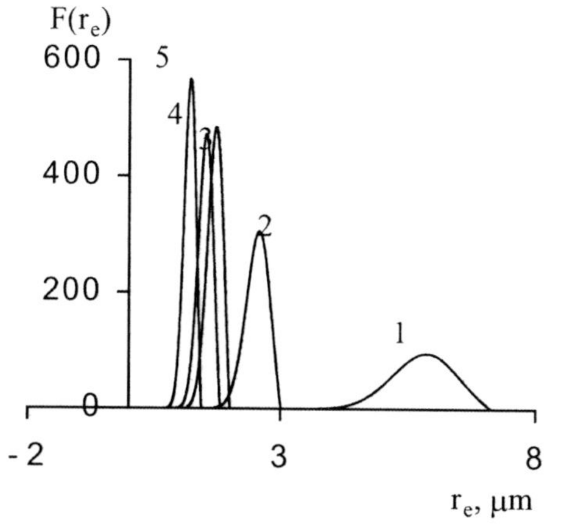

Fig. 3. The differential curves of sedimentation of particles $BaSO_4$. Speed of agitating (variant 2): 0 (1); 300 (2); 1500 (3); 2500 (4) rpm; 5 - tubular turbulent device.

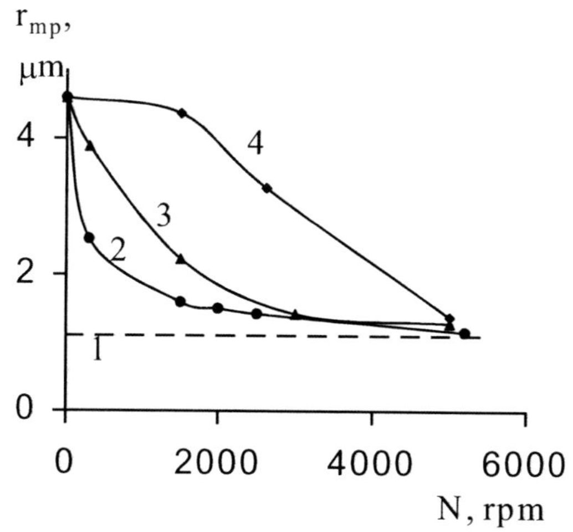

Fig. 4. Dependence of most probable radius r_{mp} of particles $BaSO_4$ on speed of agitating N. The tubular turbulent device (1); variant 1 (2) variants 2 (3, 4); after pre-award formation in absence of agitating and endurance 90 min. (4).

It is visible (fig. 4), that in case of use for suspensions of stirred tank rector (variants 1 and 2) the place decrease r_{mp} and an increase of speed agitation N (from r_{mp} = 4.6 μm in absence of agitation up to r_{mp} = 1.2 μm at N = 5000 rpm). In this case reaction mixture on conditions of experiment rather diluted and increase of turbulent agitation does not mean an increase mass transfer and, as a consequence to increase growth rate of crystals, and in result their size. The appreciable reduction of particles of the dispersed phase radius occurs for account of high speed of micromixing, which intensity is increased with the dissipation of specific drop energy of turbulence ε with increase of speed agitating [5]. At high speed of micromixing the speed nucleation/growth rate of crystals ratio is increased, that results in the increase of number nucleuses of a new phase and reduction r_{mp}. It is also observed in case of use of a mixing baffle (variant 2), which increase intensity of turbulent mixture in the stirred tank reactor. In this case, the relative speed of reduction r_{mp} with the increase of revolutions of agitator, N is higher in comparison with the first variant of a way synthesis. Pays attention that r_{mp} with increase of speed agitating in stirred tank reactor (flask) down to limiting possible in laboratory conditions (N ≈ 5000 rpm), both on variant 1, and on variant 2, aspire to the size of particles of suspension $r_{mp} \approx 1.1$ μm, which obtained in the tubular turbulent device where the high level of turbulent agitating is created extremely for account of geometry of the channel and hydrodynamical energy of flow [1]. With the purpose of an estimation of degree polydispertion of obtained suspensions on the sedimentation curve (fig. 2) the radiuses of particles expected average weigh r_w and average numerical r_n is calculate as:

$$r_w / r_n = \frac{\sum r_i n_{wi}}{\sum r_i n_{ni}}, \tag{4}$$

Where n_{wi}, n_{ni} - weight and numerical share of particles with radius r_i. For homogeneous suspension factor of polydispersity $r_w/r_n = 1$. An increase of intensity agitation during formation of solids in system on variant 1 results in a decrease of polydispersity factor from $r_w/r_n = 2.8$ in absence of agitation up to $r_w/r_n = 1.55$ at N = 5000 rpm (fig. 5). The uniformity of suspension obtained on variant 2, practically does not change with increase of intensity of turbulent agitation in the order $r_w/r_n \approx 1.77$. The fast reaction with precipitation in the tubular turbulent device, allows to form suspensions with rather high uniformity $r_w/r_n \approx 1.48$ (fig. 5). Naturally, during work was presented expedient to study the characteristics of suspensions exposed to hydrodynamical influence not at the moment of their formation, and after pre-award endurance them in absence of agitating. For this purpose the interaction of feeding reagents (water solutions H_2SO_4 and $BaCl_2$) was carried out in absence of agitation with endurance during 90 min. and subsequent influence on obtained suspensions of turbulent agitation on variant 2 and with use of the tubular turbulent device. It has appeared, that in this case in process of endurance of suspension the increase of average radius r_{mp} of particles does not occur (fig. 4, curve 4). An increase of intensity of turbulent agitation of a mix containing previously formed suspension, is not observed falloffs of radius solids obtained at absence of agitation (fig. 4, curve 4), in comparison with suspension obtained at different speed of rotation of agitator (fig. 4, curve 2 and 3). It also was shown in [6]. However, particle with r_{mp} = 1.2 μm are formed at intensive hydrodynamical influence in the stirred tank reactor (at N = 5000 rpm). The use of the tubular turbulent device for influence on the suspension, sustained during 90 min., allows to obtain dispersible systems with the even smaller size of particles ($r_{mp} \approx 1.1$ μm) (fig. 4, curve 1). Similar results are observed for the coefficient of polydispersity r_w/r_n of obtained suspensions (fig. 5). The use of the tubular turbulent device, both at the moment of synthesis suspension, and at influence on the dispersible system, previously obtained and sustained during 90 min., results in reception of rather uniform particles with coefficient of polydispersity $r_w/r_n \approx 1.48$ (fig. 5, curve 1).

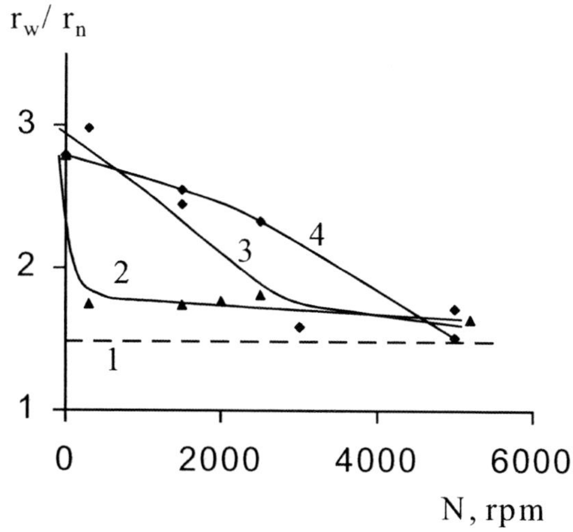

Fig. 5. Dependence the coefficient of polydispersity r_w/r_n of particles $BaSO_4$ by the speed of agitating N. The tubular turbulent device (1); variant 1 (2); variant 2 (3, 4); after pre-award formation in absence of agitating and endurance 90 min. (4).

Thus, the expediency of use of small-sized high capacity tubular turbulent devices for realization of fast chemical reactions with obtaining homogeneous high dispersity of suspensions is revealed. In this case, the residence time of a reaction mixture in the device is 2-3 sec, that reduces probability of passing of secondary processes. These devices are effective not only at synthesis, but also at processing already previously of obtained suspensions. As a consequence, it is necessary to use tubular turbulent devices for obtaining of Ti-Al, V-Al, Nd-Al and other high-activity microheterogeneous catalytic systems Ziegler-Natta in processes of (co)polymerisation of olefines and dienes (cis-1,4-isoprene, ethylenpropylene, cis-1,4-butadiene etc. rubbers), antiagglomerates (stearates Me^{2+}) for synthetic rubbers, pigments, in particular, on a basis TiO_2 and others homogeneous high dispersity of suspensions formed at interaction of homogeneous liquid flows. Apparently, that the devices of this type on efficiency do not concede to use for synthesis of suspensions pump or another the equipment.

CONCLUSIONS

The use of the divergent-convergent tubular turbulent device, both at the moment of synthesis, and at processing ready suspension $BaSO_4$, allows to obtain suspensions with average equivalent radius of particles of a dispersed phase $r_{mp} \approx 1.1$ μm and coefficient of polydispersity $r_w/r_n \approx 1.48$, that is achieved in the stirred tank reactors with a mechanical agitator, only at speed of agitating more than 2000 rpm.

ACKNOWLEDGMENTS

The work is executed at financial support of the Russian fund of fundamental researches (02-01-97913).

REFERENCES

[1] Minsker, K.S., Zakharov, V.P., Berlin, Al.Al., *Theoretical Foundations of chemical Engineering*, *35:2* 162-167 (2001).

[2] Berlin, Al. Al., Minsker, K.S., *Science to production*, *3* 7-12 (2002).

[3] Kasatkin A.G., *Basic processes and devices of chemical technology*. Moscow: Chimiya, 1971, 784 p.

[4] Frolov Yu.G. and Grodskii A.S., *Laboratory works and tasks on colloid chemistry*. Moscow: Chimiya, 1986, 216 p.

[5] Kim, W.S. and Tarbell J.M., *Chemical Engineering communications*, *176* 89-113 (1999).

[6] Sung, M.H., Choi, I.S., Kim, J.S., Kim, W.S. *Chemical Engineering Science*, *55* 2173-2184 (2000).

In: Chemical Reactions in Liguid and Solid Phase
Editors: G. E. Zaikov and A. Jiménez, pp. 69-118 ISBN 1-59033-854-5

Chapter 4

PHENOLIC RESINS AND THEIR DEGRADATION BEHAVIOUR

L. B. Manfredi and A. Vázquez[*]
Research Institute of Material Science and Technology (INTEMA) – University of Mar del Plata, Juan B. Justo 4302, 7600, Mar del Plata, Argentina
e-mail: anvazque@fi.mdp.edu.ar

ABSTRACT

The first part of this review article deals with the synthesis, characterisation and kinetic of polymerisation of the phenolic resins. Phenolic resins are polycondensation products of phenol and formaldehyde and are classified as novolaks and resols, depending on the type of catalyst and the molar ratio of reactants. Novolaks are catalysed by acids with a formaldehyde to phenol molar ratio (F/Ph) in the 0.5-0.8 range. Instead, resols are base-catalysed and F/Ph ranges from 1 to 3. The influence of the synthesis conditions such as: type and concentration of catalyst, pH, temperature and formaldehyde to phenol molar ratio have been thoroughly studied by many research groups using all kind of techniques. Then, the relationship between the synthesis conditions and the structure and mechanical properties of the cured resin was analysed. Physical and end-use properties are largely affected by the method of polymerisation. Novolak resins have low cross-linking and process like thermoplastics. Resols are highly crosslinked, versatile and stabile resins. The wide range of applications of phenolic resins includes coatings, adhesives, thermal insulators and matrix of composites. This is due to its low cost, dimensional stability, excellent chemical, moisture and heat resistant, and good fire-smoke-toxicity characteristics. Processing of phenolic-based materials requires a careful application of processing conditions that need an accurate knowledge of the polymerisation kinetics of the resin as a function of the temperature. So, kinetic models have been developed and applied to describe the chemical reactions of the phenolic resins.
Further this article gives a survey about thermal degradation and fire resistant of the resins. The thermal degradation of phenolic resins have been widely reported, mainly focused on the chemistry of the principal volatile products and on the pyrolisis behaviour.

[*] Correspondence Author

Different mechanisms for the degradation process and for different ranges of temperature were proposed. The thermal degradation occurs by scission of polymer chains forming methane, phenol and phenolic compounds, and oxidative degradation leads to carbon monoxide, carbon dioxide, benzaldehyde, benzene and toluene. Phenolic resins have a great tendency to char thus reducing the formation of volatiles. Thermal stability of resol resins with different F/Ph molar ratio and the relationship between the chemical structure and their thermal and fire resistance was studied. Distinct kinetic models for degradation reactions proposed in the literature were applied to evaluate the activation energies for the degradation process of the phenolic resins.
Nowadays, due to the toxicity of phenol and formaldehyde, there is an interest in replace them by products obtained from renewable resources. The properties and the resistance to thermal degradation of these natural resins were compared with the traditional phenolic resins.

Keywords: phenolic resins, synthesis, kinetic model, characterisation, dynamic-mechanical properties, thermal degradation, fire resistance.

1. Introduction

Phenol-formaldehyde resins are one of the first developed thermosetting resins. They are widely used in industries for the manufacture of coatings, adhesives, electrical insulation parts and composites, owing to its low cost, dimensional stability, chemical, moisture and fire resistance and high tensile strength (Knop and Pilato [1]). The brittle nature however, has significantly limited its usage. Many efforts are being delved to improve the structural and decorative properties through blending the resin with other polymers (Wu et al. [2]). The hydroxyl group in phenolic provides the hydrogen donor forming hydrogen bonding with polymers containing hydroxyl, carbonyl, ether or ether proton - accepting segments. Several such polymers have been successfully used as modifiers to tailor the phenolic physical properties (Yang et al. [3]; Matsumoto et al. [4]; Matsumoto et al. [5], [6]; Wu et al. [7]) and this inter-polymer association are considered to play a key role in polymer compatibility between phenolic and the modifiers. Relationships between the chemical network structures and the physical and flame performance properties of novolak phenolic - epoxy materials have been studied recently by Tyberg et al. [8]. The Tg decreased, and toughness increased, with the increase of the novolak content in the network. These results could be due to the decrease in the network densities of the compounds. Analysis of the cooperativity of the networks suggested a crossover in properties from two competing factors, network density and intermolecular forces (hydrogen bonding). In addition, an increase in novolak content improves the flame retardance. Thus, it can be possible to obtain a void-free system with both favourable mechanical properties and flame retardance by controlling the phenol to epoxy ratio.

In general, thermosets polymers have the advantages of chemical resistance and thermal stability. But, due to their low rigidity and mechanical resistance, they are used as matrix for the fabrication of composite materials. The combination of a thermoset matrix with glass fibre is also used in a variety of applications from transportation, construction, marine, corrosion resistant equipment, aircraft to sports products. High performance and advanced composite are fibre-reinforced composite that combine the lightweight with the good structural properties.

Phenolic resins are used as matrix in different composite market segments depending on the end use applications. The combination of attractive fire – smoke – toxicity characteristics, reasonably priced resin and emerging driven applications have been the motivating factors for the resurgence of interest in phenolic resins for fibre reinforced plastics.

Phenolic resins composite materials could be obtained by different types of processability and could be combined with a varied range of fibres (Gardziella et al. [9]). The pultrusion is one of the segments that is increasing faster in the composite industry (Wilder [10]).

Phenolic resins are superior to epoxy, vinyl ester or polyester resins under fire conditions. These fire conditions allow the phenolics to be used as matrix in a fibre reinforced composites for demanding critical applications, e.g., where the safety of the passengers is important during egress in case of fire (aircraft interiors, mass transportation, and marine/offshore areas). The carbonisation and char insulating characteristics of the phenolics by controlled pyrolisis with the chemical resistance and strength retention allow their used in composite materials in oilfield applications (pipe, platform decking).

Recently, investigations made by the National Bureau of Standards (NBS) lead to improve the security in the transportation systems and buildings from the use of fire resistant panels, design of fire resistant interiors of vehicles and furniture. The European market is the leader in the topic of fire resistant materials and in the last years, the train industry was the principle investors in security against fire. The security fire standards applied in the eurotunel (that link France with England) were so strict that the phenolic resin composites were the ones approved to be utilised (Brown et al. [11]).

Even though the phenolic resins have good fire properties, they are difficult to process. Because of this, the development of phenolic resins with better properties to be used in glass fibre reinforced composites by means of superficial treatment of the fibre have acquired a new significance (Gardziella et al. [9]). Recently, phenolic resins less brittle than the traditional ones were developed and have maintained their properties of flame retardant and low smoke emission. These resols were cured at low temperatures with the addition of acid catalyst (St. John and Brown [12]).

2. Chemistry of Phenolic Resins: Reaction Mechanism of Phenolic Resins and their Relationship with Reaction Conditions and Chemical Structure

Phenolic resins have been in commercial use longer than any other synthetic polymer except for cellulose nitrate. Martin [13], Knop and Scheib [14], Knop and Pilato [1], Gardziella et al. [9] are general references on this topic. Phenolic resins are based almost exclusively on the reaction products of phenol and formaldehyde by step-grow polymerisation. Substituted phenols of lower functionality are used to incorporate specific properties in the resin (adhesives, coatings).

Depending on the formaldehyde to phenol molar ratio (F/Ph) and catalyst, novolak or resol resins can be formed. Novolak resin is formed in the presence of acid catalyst when F/Ph<1. The methylol derivates undergo exothermic condensation with phenol to form a linear polymer, where *ortho* and *para* links occur randomly (Aranguren et al. [15]). Resol formation takes place in the presence of alkaline catalysts when F/Ph>1. The methylol phenols can condense either through methylene linkages or through methylene ether linkages. Resol resin is similar to novolak but it contains reactive methylol groups that can condense to a network structure. So, it is possible to distinguish the following reaction steps: the formaldehyde addition to phenol, the prepolymer formation and the crosslinking reaction, at temperatures above 100°C.

The final structural and mechanical properties of the phenolic resins depend on many factors that are involved in the synthesis of the prepolymer: initial formaldehyde to phenol molar ratio, temperature and time of condensation, pH and nature and amount of catalyst.

It is important to determine the phenol and formaldehyde functionalities because the type of resin obtained is related with the F/Ph molar ratio. Drumm and Le Blanc [16] experimentally determined the phenol average functionality in novolaks as equal to 2.31. More recently, Shipp and Solomon [17] reported that it is in the range between 1.49-1.72. However, the phenol functionality in resols is more complex and it could be ≤ 3 (Gardziella et al. [9]). Moragues [18], reported a phenol functionality in resols of 2.5, considering the formaldehyde functionality as equal to 2.0. This value of "real" or "effective" functionality could be attributed to structural effect between the rigid chains of the polymer that limit the mobility and to the different reactivity between the internal and external sites (at the end of the chain) of the polymeric chain.

2.1. Phenol-Formaldehyde Reaction under Acidic Conditions

The two possible reactions are addition and condensation.

2.1.1) Formaldehyde Addition to Phenol and Prepolymer Formation

The addition reaction could be produced in strong or weak acidic conditions

2.1.1.1) Addition Reactions under Strong Acidic Medium

The formaldehyde to phenol addition is an electrophilic aromatic substitution in acidic as well as in alkaline conditions. The pathway of reaction is much more complicated with resols because of solvent interactions and inter- and intramolecular hydrogen bond formation. This is supported by the abnormal and wide variation in *ortho/para* ratio.

In aqueous medium a very fast acid and base catalysed hydratation reaction of formaldehyde to methyleneglycol occurs, eq (1)

$$CH_2{=}O + H_2O \rightleftharpoons HO-CH_2-OH \text{ , } K= 7.14 * 10^{13} \tag{1}$$

The concentration of non-hydrated formaldehyde is less than 0.01%. In aqueous solutions, methyleneglycol is present as a low molecular condensation polymer. It is present as a monomer only in a very dilute aqueous formaldehyde solutions (1-2%). The overall reaction rates for a phenolic resin depends on the depolimerisation of the polyoxymethylene. Under acidic conditions, methylene glycol is converted to the corresponding hydrated carbonium ion:

$$HOCH_2OH + H^+ \rightleftharpoons {}^+CH_2OH + H_2O \quad (2)$$

Then, the hydroxymethylene carbonium ion adds to the *ortho* and *para* positions of phenol:

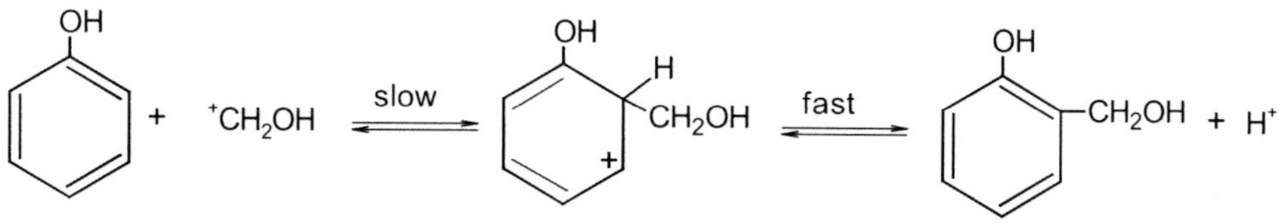

(3)

This is a relatively slow step and it determines the rate of the reaction. There is no substitution in the *meta* position of the phenol. Then, the methylol group result in benzylic carbonium ions:

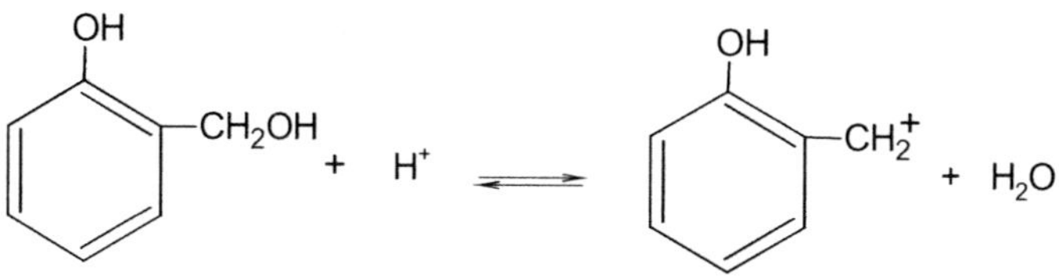

(4)

In the next step, the benzylic carbonium ion reacts very fast with free *ortho* and *para* positions of phenol yielding methylene-linked bisphenols:

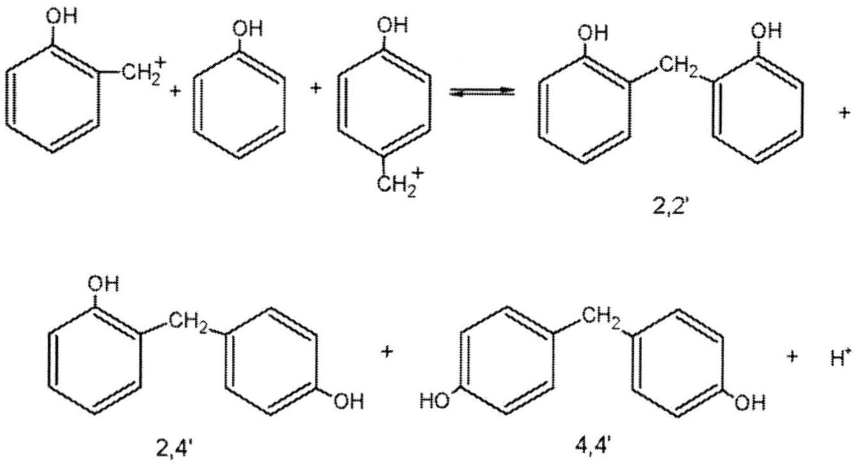

(5)

Methylol substitution and methylene bridge formation both occur preferably at the *para* position.

The reaction continues and forms the novolak polymer with a molecular weight of 500-5000 and a glass transition temperature (Tg) of 45-70°C.

The typical acid catalysts used in the synthesis of novolak resins are: sulfuric acid, sulfonic acid, oxalic acid, p-toluene sulfonic acid or occasionally phosphoric acid.

The novolak molecules with a molecular weight up to 1000 are essentially linear because of the much lower reactivity of doubly reacted phenolic units. The novolaks with a higher molecular weight are branched because the low concentration of end groups and unreacted phenol.

The acid catalyst and reaction conditions affect the resin structure and final properties (Kopf and Little [19]).

The kinetics of the reaction under acidic condition was widely studied (De Jong and De Jonge [20]; Von Euler [21]; Jones [22]; Bassow [23]). They found that the overall reaction rate followed a second order reaction kinetics. The rate of formation of the benzyl alcohol intermediate is 5-10 times lower than the rate to form the methylene-linked bisphenol. The reaction rate is proportional to the concentration of the ion H and the activation energy increases with increasing pH. In general, a second formaldehyde addition does not occur at F/Ph molar ratios usually used.

2.1.1.2) Addition Reactions under Weak Acidic Medium

In weak acidic medium and by use of specific catalyst, high o-o' resins can be obtained. According to the F/Ph molar ratio, the high o-o' resins can be solid novolaks or liquid resols. Both of them are important in commercial applications such as moulding compounds, foundry and Reaction Injection Moulding (RIM) or Reaction Transfer Moulding (RTM).

Novolaks could be obtained in the range of pH=4-7 and using divalent metal catalysts. Zinc and calcium acetates are the most widely used salts (Bender [24]). There is an *ortho* directing effect with the formation of a chelate between the phenolic ring and the metal ion:

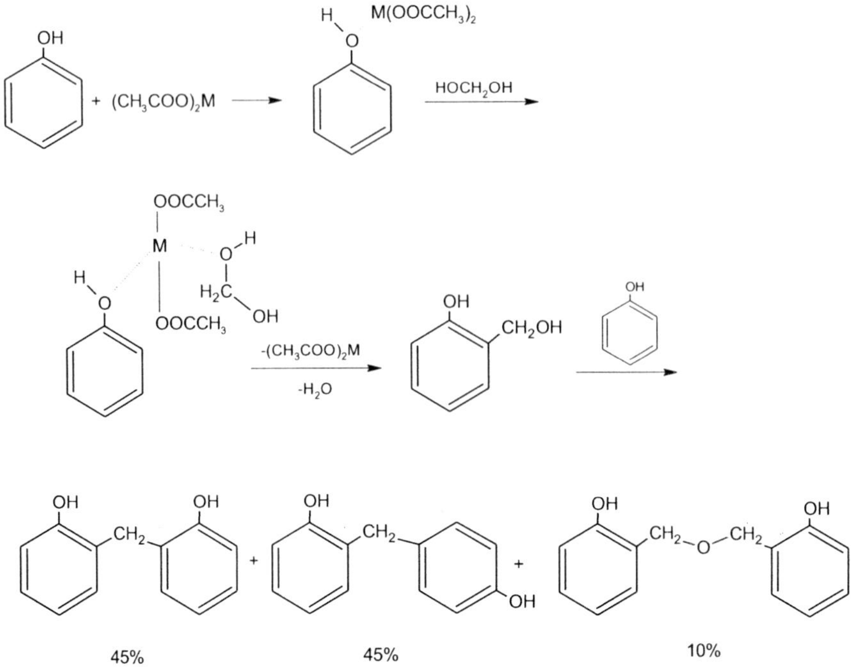

(6)

The first step gives a methylol phenol, and the second step gives an equal quantity of 2,2'- and 2,4' diphenylmethylene bridges. The ether bridges are present in less quantity. To obtain high o-o' novolaks it is necessary to lead the reaction in the same conditions but with a large excess of phenol (Knop and Pilato [1]).

Infrared spectroscopy (IR) and Nuclear Magnetic Resonance (NMR) (Higginbottom [25]) were used to characterize the o-o' structure of the o-o' novolaks:

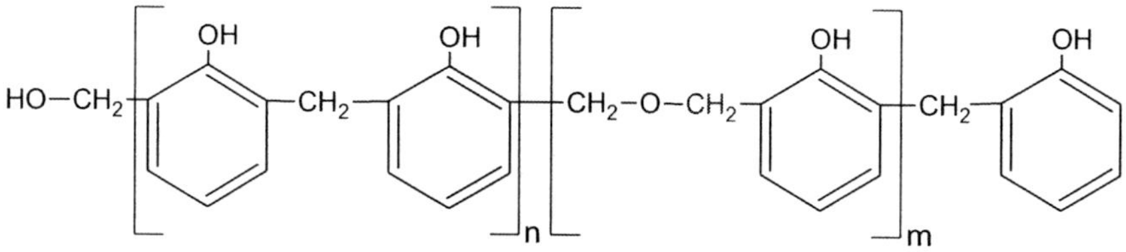

(7)

Resols are prepared at intermediate pH, using divalent salts as catalysts and in an azeotropic solvent such as xylene or toluene. The temperature was maintained at 80-90°C during 5-8 h. The Ph/F molar ratio was 1/1.5-1.8.

The structure proposed for this resin was:

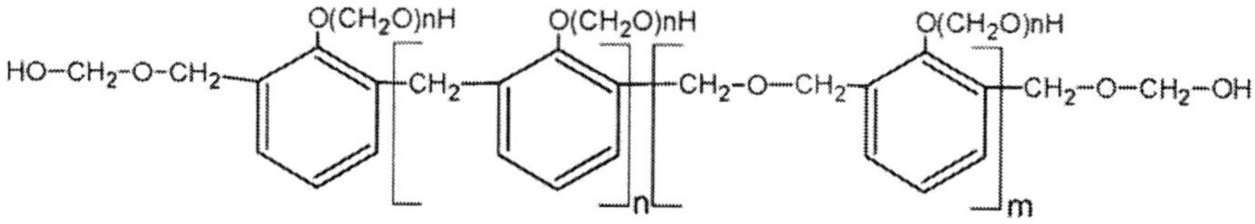

(8)

These resins have low viscosity, high solids content and better storage stability than the conventional resols.

2.1.2) Crosslinking Reactions of Novolaks

After that the prepolymer is obtained, in order to have solid part, the crossliking of the phenolic prepolymers resins is performed.

Hexamethylenetetramine (HMTA) is the most used curing agent for novolak resins. The amount used of HMTA is between 5 and 15%. From the ratio of the formaldehyde to phenol used in the synthesis of novolaks it was possible to determine to a great extent the properties of the cured resins. The curing reaction of the novolak/HMTA was studied by Hatfield and Maciel ([26]) who proposed the following two stages mechanism:

(1) the formation of intermediates, such as benzoxazines and benzylamines. The amount of vacant *ortho* and *para* sites affect the relative amount of each intermediate.
(2) decomposition, oxidation, and/or further reactions of these initial intermediates. Methylene bridges were formed between the phenolic rings and amines, amides/imides, imines, methylphenol, benzaldehyde and other trace materials.

The pH of the novolak/HMTA composition affects the reaction. Novolaks are moderately acidic and HMTA is basic. At lower pH and using a low content of HMTA, aprox. 6%, the reaction rate is faster to initially form cured intermediates. Then the decomposition and further reactions of initial intermediates occurs. The intermediates with *ortho*-methylene bridges are more stable than the *para*-bridged intermediates due to the formation of intramolecular hydrogen bond interactions. The *ortho*-intermediates stabilize forming six-member ring structures.

The ratio of *ortho/para* sites of the starting resin and the amount of initial HMTA affect the final chemical structure of the cured novolaks. Using a high content of HMTA and a novolak with a high amount of *ortho* sites, it is possible to obtain a crosslinked structure with a high content of nitrogen. On the other hand, using a low content of HMTA and a novolak with a high amount of free *para* sites, it is possible to obtain a crosslinked structure but with a low nitrogen content (Gardziella et al. [9])

2.2. Phenol-Formaldehyde Reaction under Alkaline Conditions

The reactions that occur under these conditions are: formaldehyde to phenol addition, condensation and crosslinking reaction in acidic medium or with temperature

2.2.1) Formaldehyde to Phenol Addition

Phenol reacts quickly with alkali hydroxide to form the phenolate anion, which stabilizes by resonance. The delocalisation of the negative charge is in the *ortho* and *para* positions, eq (9)

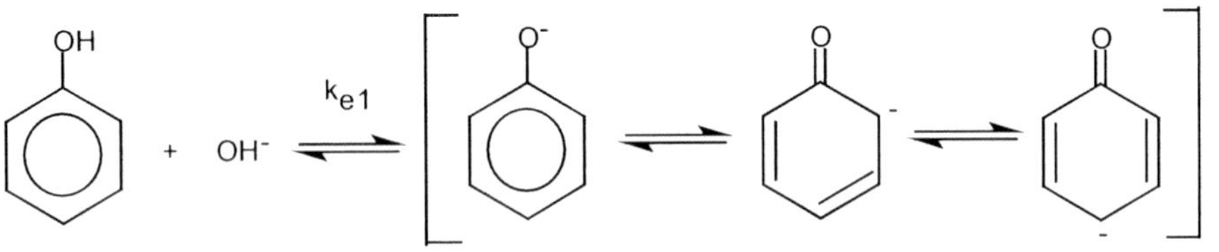

(9)

So, the formaldehyde addition to phenol occurs almost exclusively in *ortho* and *para* positions.

As it was previously explained, in a base catalysed medium a hidratation reaction of formaldehyde to methyleneglycol occurs, eq (1). The formaldehyde concentration, in the form of $CH_2=O$, is less than 300 ppm so it is not probably involved. Its concentration is too low to explain the reaction rates.

Different mechanisms have been proposed to explain the formaldehyde to phenol addition reaction (Kopf and Little [19]). One of them propose the presence of hemiformals that rearrange to form hidroxymethylphenoxide:

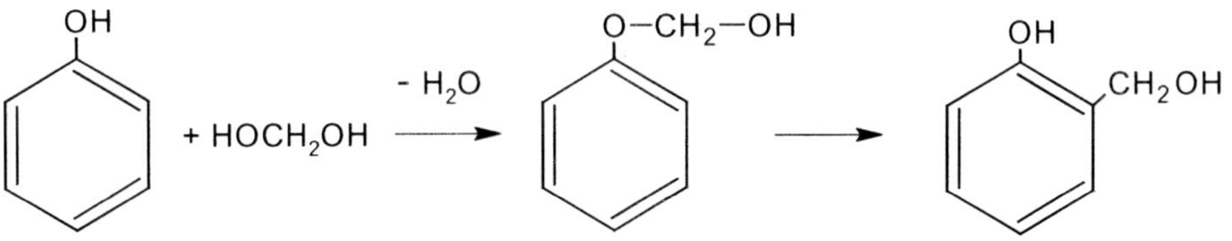

(10)

Another mechanism involves the methylene glycolate anion $HOCH_2O^-$. But probably it is present in a very low concentration because the methyleneglycol is a much weaker acid than the phenol. The most likely mechanism is the one that involves a nucleophilic displacement by the phenoxide on the methylene glycol, with the hydroxyl as a leaving group. Then, the quinone intermediate is rapidly converted to the phenoxide (Zsavitsas and Beaulieu [27]), as it is shown in eq (11 and 12), for the *ortho* and *para* positions:

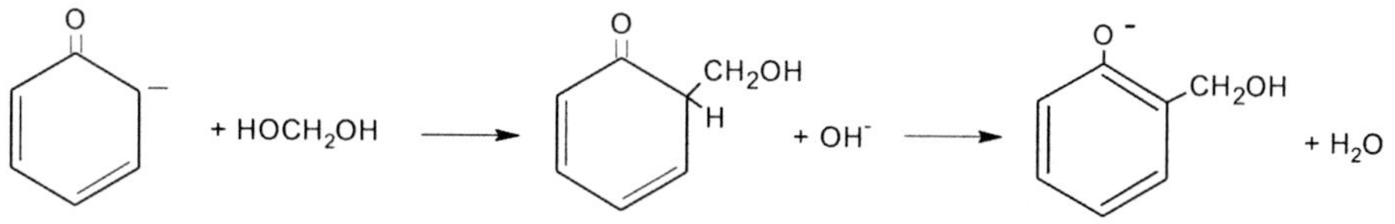

(11)

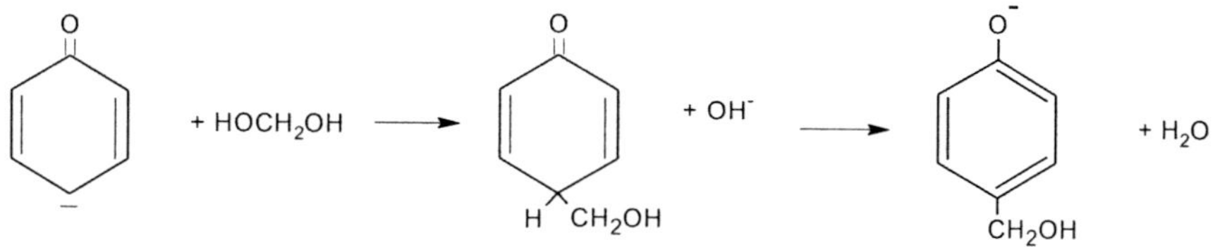

(12)

However, there have been some evidences that indicate that the transition state is not as simple as indicated in eq (11 and 12). The most important evidences are that the ratio of *ortho* to *para* substitution depends on the nature of the cation of the catalyst and the pH value (Peer [28]; Peer [29]).

The alkaline catalyst usually used in the resols synthesis are NaOH, $Ca(OH)_2$ and $Ba(OH)_2$, but NH_3, hexamethylenetetramine (HMTA) and tertiary amines are also used (Knop and Pilato [1]). The *para* substitution is favoured at higher pH and with Na^+ and K^+, whereas *ortho* substitution is favoured at lower pH and with divalent cations (Ba<Sr<Ca<Mg). The chelating strength of the cation favours the *ortho* substitution. Peer proposed the formation of these chelates for transition metal hydroxides like Fe, Cu, Cr, Ni, Co, Mn and Zn, and with boric acid.

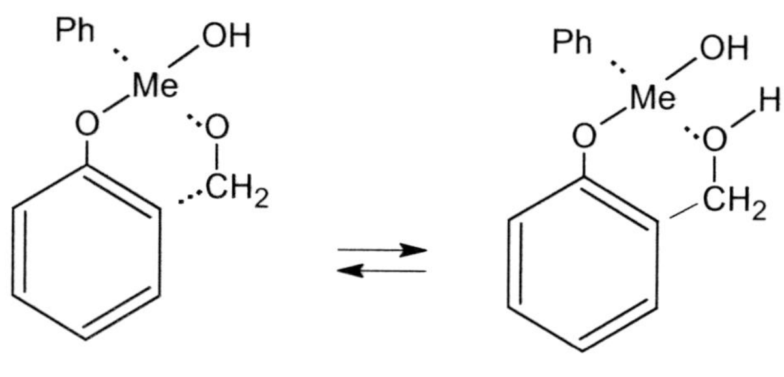

(13)

The monomethylolated derivatives continue reacting with formaldehyde to form the di- and trimethylolated compounds.

In addition the reaction between the hydroxymethylphenols and the methylene glycol must be taken into account. The reaction product is:

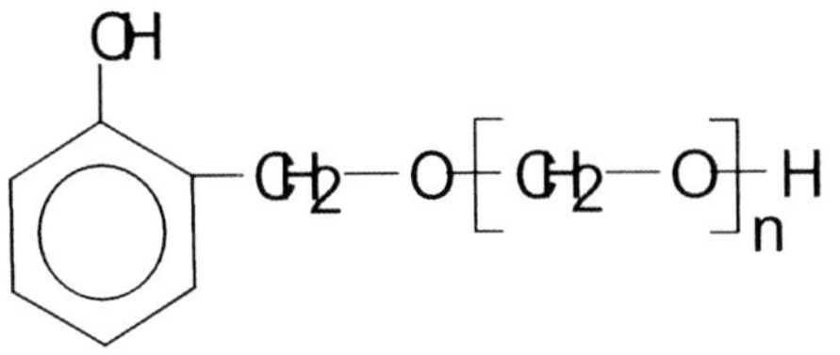

The peaks for n = 0, 1, 2 and 3 were identified by NMR (Knop and Pilato [1]). Zsavitsas and Beaulieu ([27]) explained the reduction in the reaction rate by the consumption of formaldehyde to form hemiformals.

When the catalyst used in the synthesis is ammonia or amines (except tertiary amines), the resols obtained have different chemical and physical properties. The reaction of phenol, formaldehyde and amine is an aminoalkylation:

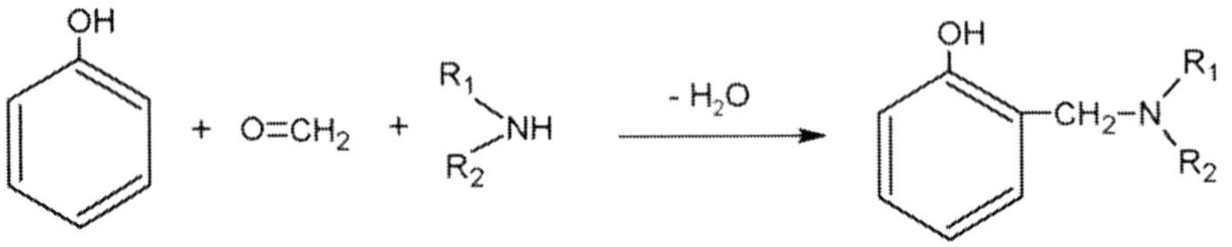

(14)

When HMTA is the amine used, the final resol obtained is the same but the initial reaction step differs. HMTA is the reaction product of formaldehyde with ammonia or aliphatic amines. The simultaneous reaction that occurs is the formation of methylol derivatives from phenol and formaldehyde.

The resins catalysed with amines has higher molecular weight, less free phenol, lower water solubility and higher Tg than one resin catalysed with NaOH (Kopf and Little, [19])

2.2.2) Condensation Reactions

At lower temperatures (60°C) only the addition of formaldehyde to phenol occurs, and at temperatures between 60 and 100°C, the following condensation reactions take place:

- between two methylolphenols to form methylene bridges, water and formaldehyde
- between two methylolphenols to form ether bridges and water
- between methylolphenols and phenol or methylolphenols free sites (2, 4, 6) to form methylene bridges and water
- as it is shown in eq (15 a, b, c and d)

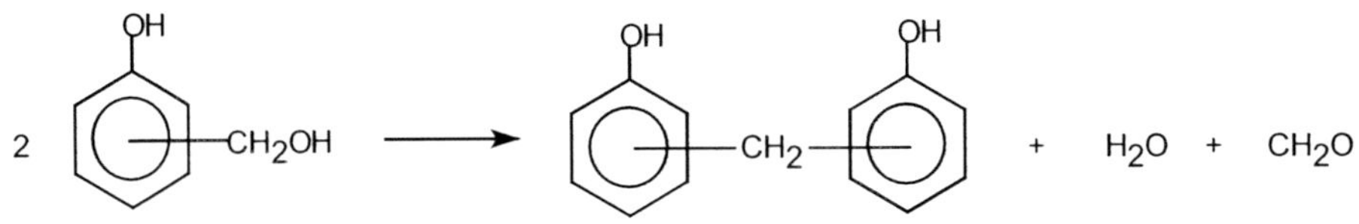

a)

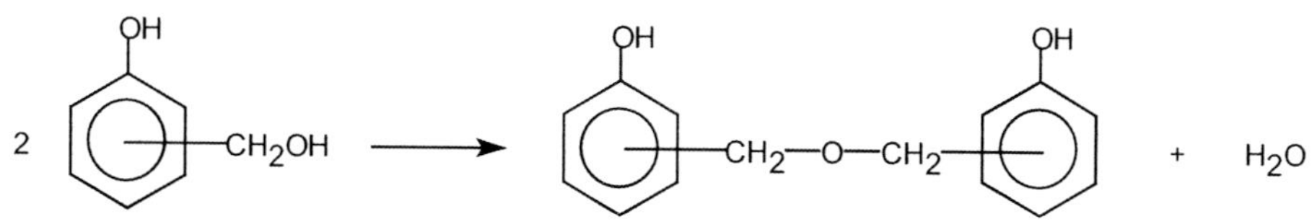

b)

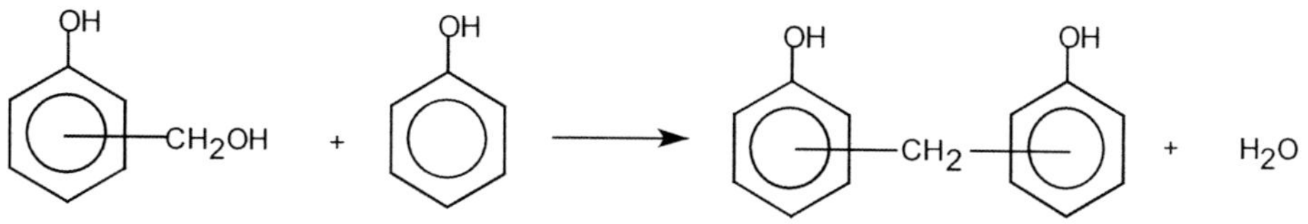

c)

+ H_2O (15)

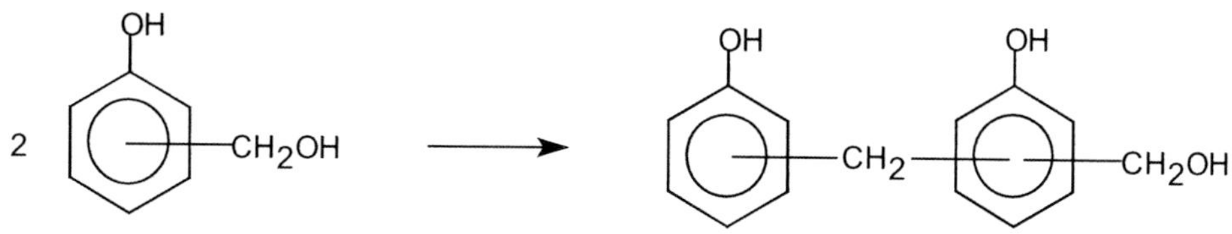

(d)

The methylene bridges are formed in an alkaline medium and at temperatures higher than 130-150°C. On the other hand, in a neutral or slightly acid medium and at lower temperatures, the ether bridges are formed (Maciel et al. [30]; So and Rudin [31]; Grenier-Loustalot et al. [32]).

Grenier-Loustalot et al. ([32]-[35]) in a series of publications, studied the resols preparation in controlled conditions of temperature, stoichiometry, catalyst and pH by techniques like HPLC, ^{13}C NMR, FTIR, UV and chemical assays.

The formaldehyde consumption and the formation of methylolated monomers, dimers and trimers were studied. It was demonstrated that the resin structures depend on the synthesis conditions. The presence of o-p' and p-p' methylene bridges were observed; but o-o' methylene bridges were no present.

The use of different basic catalyst was studied (Grenier-Loustalot et al. [33]): Group IA hydroxides, like LiOH, NaOH, KOH; Group IIA hydroxides, like $Mg(OH)_2$, $Ca(OH)_2$ and $Ba(OH)_2$ in similar reaction conditions. The velocity of formaldehyde consumption was related with the ionic radius in each group: Li > Na > K and Mg > Ca > Ba. Comparing the different catalyst groups: Mg > Ca > Ba > Li > Na > K; it was proposed that this behaviour is related with the ionic radius of the hydrated cations.

Astarloa-Aierbe et al. ([36]-[38]) fully studied the influence of the type and amount of alkaline catalyst and pH used during the resol synthesis. The kinetic of the resol reaction catalysed with NaOH, BaOH and trietylamine, was studied by HPLC. The presence of hydroxyl groups in the catalyst gave the prepolymer similar characteristics as reddish colour, time of condensation to reach the final prefixed point, condensation products. However, the catalysis with barium hydroxide favoured larger addition of formaldehyde onto *ortho* phenolic sites with respect to that for sodium hydroxide, but lower than for triethylamine.

Therefore, the *ortho* directing properties of the catalysts studied increase in the order: triethylamine>barium hydroxide>sodium hydroxide.

Gupta and Hindersinn ([39]) studied the phenol hidroxymethylolation varying the pH value; the lower rate of reaction was obtained for pH 5. On the other hand, Kumpinsky ([40]) demonstrated that the rate of reaction for a pH value between 2 and 7 was minimum.

The average molecular weights $\overline{M_n}$, $\overline{M_w}$ and the polidispersity index ($\overline{M_w}/\overline{M_n}$) of different resols were determined (So and Rudin [31]; Kim et al. [41]; Holopainen et al. [42]). Rudin et al. ([43]) determined that the species separation in resol resins is incomplete by this technique because of aggregation between the phenol derivatives. There are hydrogen bondings between the different phenolic and alcoholic residues. The complete characterization of resols by GPC is very difficult. In the chromatogram there are overlapping peaks and it is difficult to assign a particular peak to a specific compound. Despite of these problems, GPC analysis is a useful tool to be used for qualitative analysis and for comparing a series of compounds.

2.2.3) Crosslinking Reaction under Acidic Medium or with Temperature

The curing process can be made by heating or by adding basic or acidic compounds. Special types of curing process use carboxylic acids, esthers, anhydrides, amides and carbonates.

During the curing process, the molecular weight begins to increase until it reaches the gel state. The modulus increases and the structure becomes rigid and crosslink. A phenolic resin completely cured has a high modulus, Tg and crosslinking density, and presents an excellent resistance to heat and humidity (Prime [44]).

The curing by heating is performed at temperatures between 130 and 200°C. Maciel et al. ([30]) studied the curing process of resol by NMR. They proposed different possible reactions, as shown in eq (16 a, b, c):

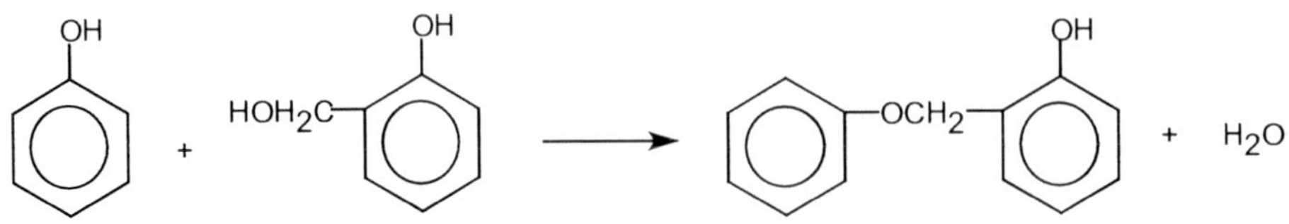

a)

(16)

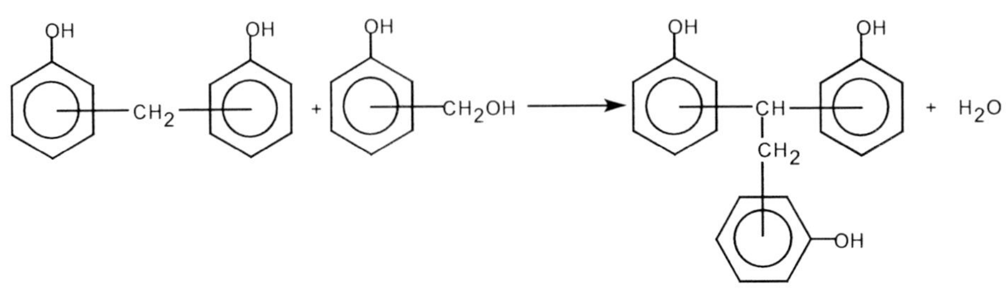

b)

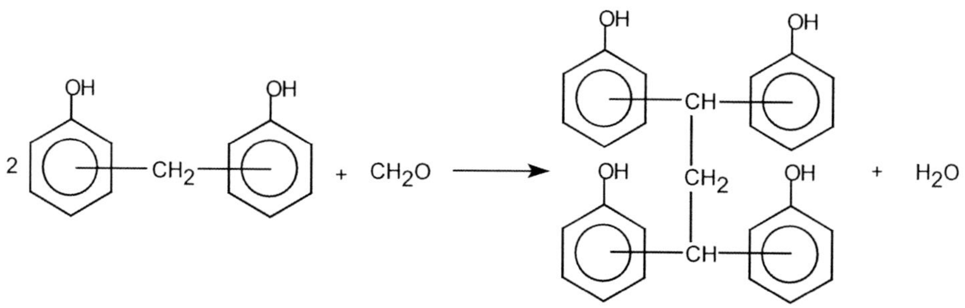

c)

The benzoquinone are the proposed intermediates compounds during the curing and condensation reactions. These intermediates are probably produced from both the methylolated phenol and the dibenzylether units, eq (17 a, b) (Kopf y Little [19]):

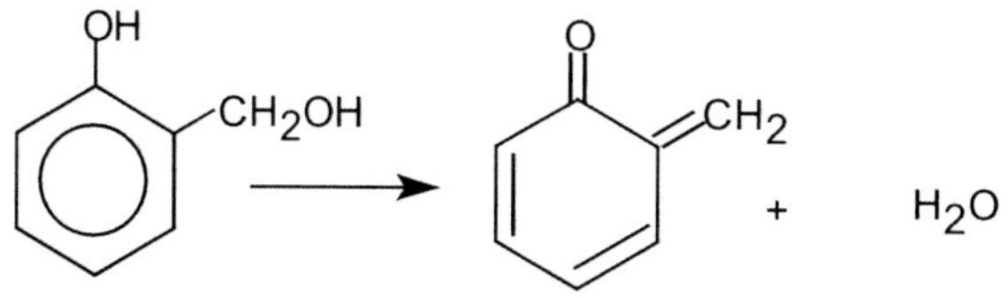

a)

(17)

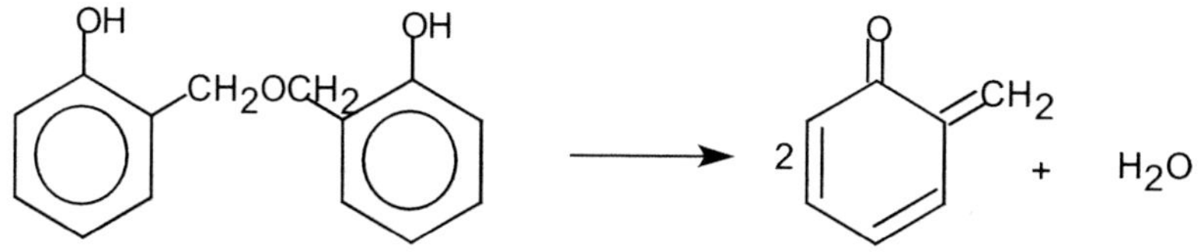

b)

Similar reactions occur for the *para* position. The quinone methide intermediates are unstable and rapidly undergo further reactions (Benson et al. [45]).

Dimerisation results in the formation of the following compounds:

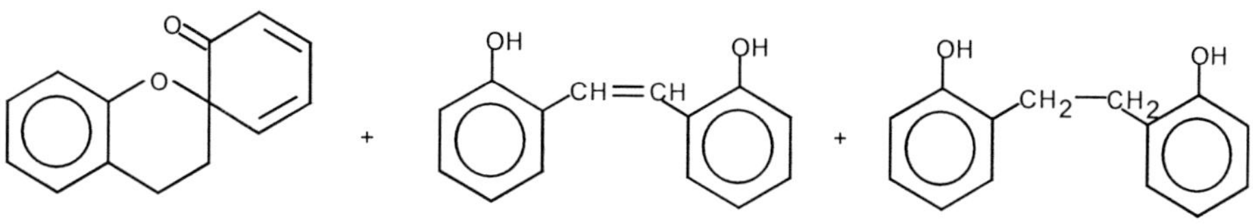

These quinone methides are responsible for the yellow and pink colour of the resins.

The curing process of the resols was studied through different techniques, as Dynamic-Mechanical Analysis (DMA) or Differential Scanning Calorimetry (DSC). DMA is a very effective method to evaluate the curing process (Myers et al. [46]; Kim et al. [47]; Follensbee et al. [48]). DSC (Christiansen and Gollob [49]; Myers et al. [45]; Wang et al. [50]) was used to determine the total heat released during the curing process and it was related with the degree of cure reached. Two peaks were identified in the DSC analysis and they were attributed to the formaldehyde addition to phenol, and to the methylol condensation, respectively. Focke et al. ([51]) and Kenny et al. ([52]) developed a kinetic model of cure, and determined the kinetic constant, activation energies and reaction orders for the curing process.

Latocha and Balcerowiak ([53]) demonstrated that during the storage at ambient temperature, the hemiformals groups in the phenolic resins, underwent slow reactions of policondensation and form methylene bridges, with elimination of water and formaldehyde. Moragues ([18]) also studied the stability of the resols. While the resols were maintained at -18°C, there was no change in the chemical structure during a time close to 230 days. However, when the resols were maintained at ambient temperature, there were changes in the viscosity and molecular weight, as it is shown in Fig 1. The concentration of species such as phenol, *ortho*-methylolphenol and *para*-methylolphenol diminished, so the condensation reaction occurs at ambient temperature.

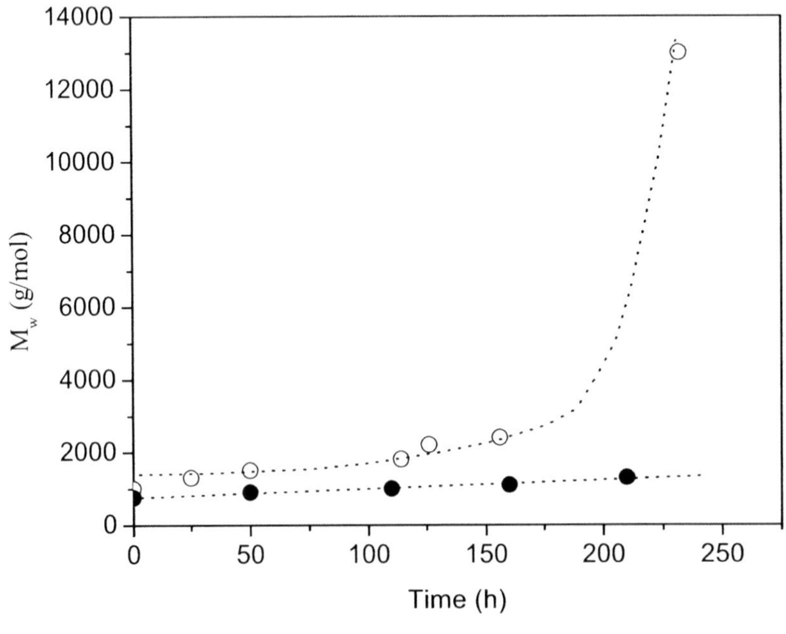

Figure 1: Molecular weight as a function of the time, for a resol maintained at: (O) ambient temperature and (●) – 18ºC.

Until this time the relationship between the synthesis conditions, the structure and mechanical properties of the cured resins has not been completely clarified yet. One of the reasons is that the resol has a complicated mechanism of reaction. So, in order to determine the influence of the chemical structure on the viscoelastic and mechanical properties, Manfredi et al. ([54]) studied a serie of resols with different F/Ph molar ratio. Six resols were synthesised with different formaldehyde to phenol molar ratio from 1.2 to 2.5, and two commercial resols with F/Ph molar ratios of 1.0 and 2.0 were also used. Firstly, these resols in solution, were characterised by means of infrared spectroscopy and chemical methods before the crosslinking reaction. Then they were cured by means of temperature up to 190°C and without catalyst. On the basis of the Grenier-Loustalot et al. ([32]; [33]) FTIR results, the different wavelengths were assigned in the spectra for the different compounds. Figure 2 shows the spectra for different uncured resols and the bands assigned to the substituted phenols.

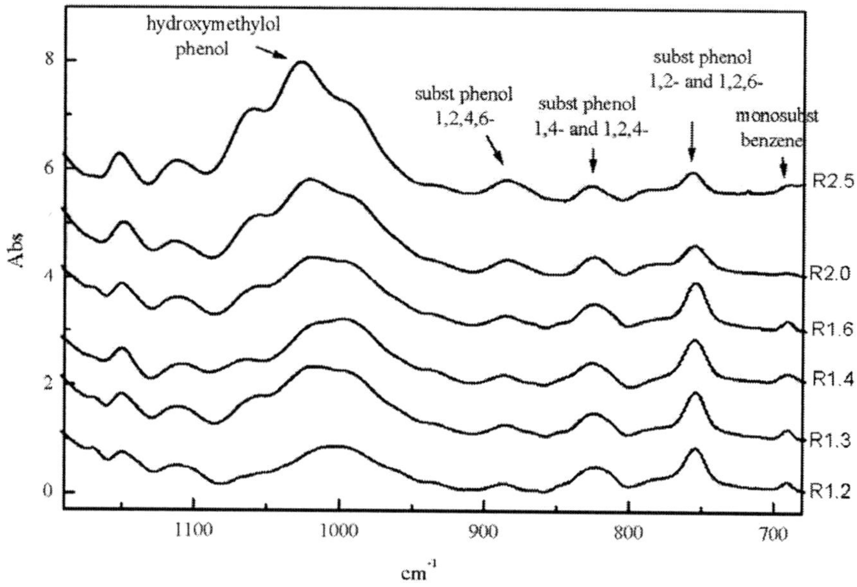

Figure 2: Spectra of uncured resols with different formaldehyde to phenol molar ratios: 1.2, 1.3, 1.4, 1.6, 2.0 and 2.5. (From Manfredi et al. [54]. Reprinted by permission of Elsevier Science)

Figure 3 shows the results for the resols with different F/Ph molar ratio. A maximum quantity of disubstituted phenol was found for a F/Ph molar ratio between 1.3 and 1.4. Figure 4 shows the spectra for the cured resols in the region where it is possible to identify the methylene bridges. The quantity of p-p' bridges is higher than the o-p' bridges, and the o-o' methylene bridges were not found in the resols (Fig.5). A higher proportion of methylene bridges were found for a F/Ph molar ratio between 1.3 and 1.4. Dynamic-mechanical analysis (DMA) of the serie of cured resols was performed and Figure 6 shows a typical curve. The measured viscoelastic properties were related with the F/Ph molar ratio. A minimum in the loss modulus value, E", as well as in the height of tanδ (=E"/E') or damping value was found for a F/Ph molar ratio between 1.3 and 1.4. In addition, the glass transition temperature, Tg, and the rubber modulus of elasticity, E'r, showed a maximum value for the same resols. It can be concluded that the real phenol functionality is close to 2.7 from the results obtained by Infrared Spectroscopy (IR) and the viscoelastic measurements for resols with different F/Ph molar ratio. Two commercial resols crosslinked without using catalyst were also characterised. In the results obtained with the commercials and synthesised resols, the main determining factor in the final properties appears to be the molar ratio of F/Ph at the beginning of the reaction.

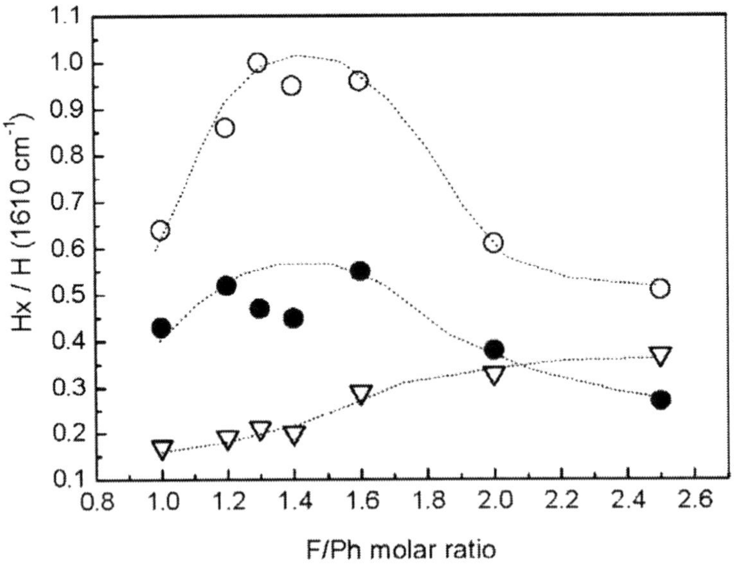

Figure 3: Changes for the uncured resols in wavelength at: (○) 756 cm^{-1} for 1,2- and 1,2,6-substituted rings; (●) 826 cm^{-1} for 1,4- and 1,2,4-substituted rings; (□) 888 cm^{-1} for 1,2,4,6-substituted rings. (From Manfredi et al. [54]. Reprinted by permission of Elsevier Science)

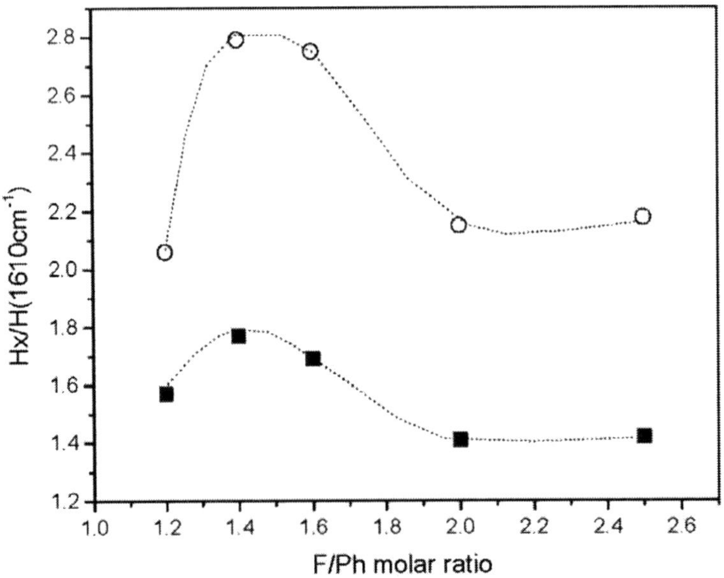

Figure 4: Spectra for fully cured resols with different formaldehyde –to-phenol molar ratios: 1.2, 1.3, 1.4, 1.6, 2.0 and 2.5. (From Manfredi et al. [54]. Reprinted by permission of Elsevier Science)

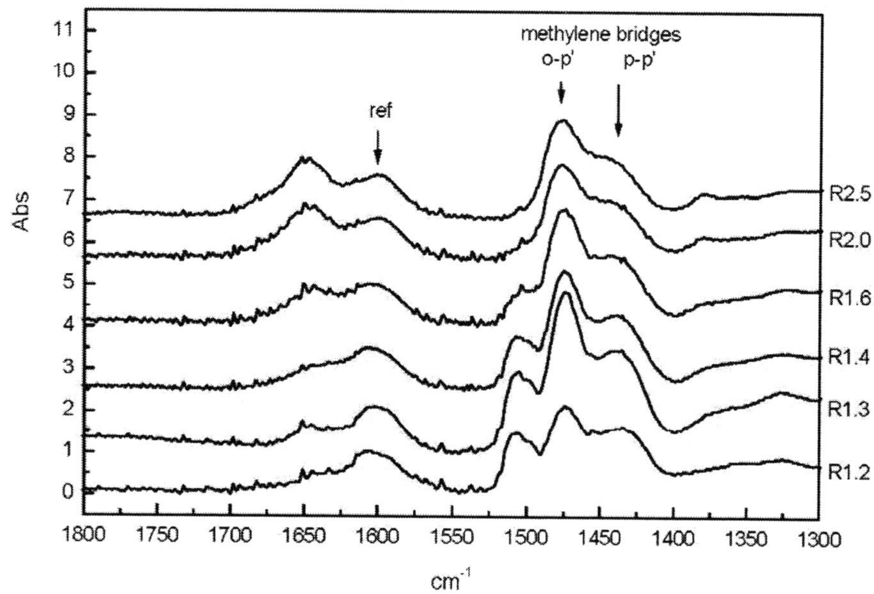

Figure 5: Changes for fully cured resols in wavelength at: (O) 1480 cm^{-1} for o-p' methylene bridges; (■) 1450 cm^{-1} for p-p' methylene bridges. (From Manfredi et al. [54]. Reprinted by permission of Elsevier Science)

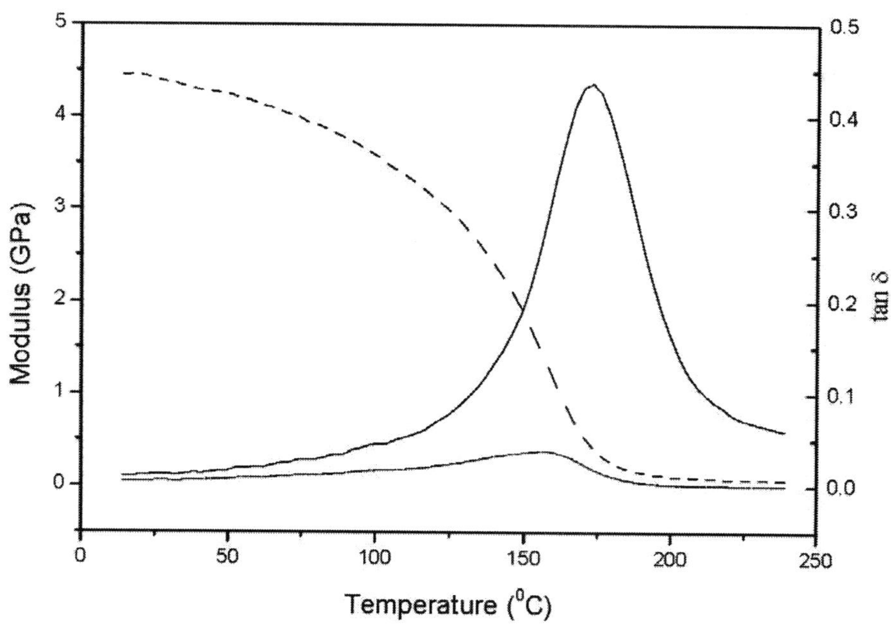

Figure 6: Typical curve of dynamical mechanical analysis for a resol showing: (----) storage modulus, E'; tanδ; loss modulus, E''.

3. Kinetics of the Reaction

The objective of a kinetic study is to determine how a reaction proceeds and predicts the formed compound in function of the time of reaction. However, the kinetic of the phenolic resins is complicated because of the influence of the reaction conditions (temperature, type and amount of catalyst, F/Ph molar ratio and time) in the results.

The kinetic and chemistry of the reaction of phenol and formaldehyde in basic medium were fully studied (Jones [55]; DeJong and DeJonge [20]; Freeman and Lewis [56]; Yeddanapalli and Kuriakose [57]; Peer [28]; Zsavitsas [58]; Zsavitsas and Beaulieu [27]; Eapen, and Yeddanapalli [59]; Zsavitsas [60]; Zsavitsas et al. [61]) and were relatively well understood. In general a second order reaction was observed, but it is a first order reaction when the ammonia is used as catalyst. The NaOH was the most used catalyst.

One of the most complete studies about the formation of the methylolated derivatives was made by Freeman and Lewis ([56]). The reaction was made at 30°C and with a molar ratio Ph/F/NaOH 1:3:1. Each individual reaction was assumed to be of the second order. The methylolated monomers formed during the addition of the formaldehyde to the phenol are shown in Figure 7. They determined that the phenol *para* position is slightly higher reactive to formaldehyde than the *ortho* position. Also, it was found that the monomethylolated derivative in the *ortho* position is produced at a higher velocity than the substituted in the *para* position, because of the presence of two *ortho* positions for each *para* position in the phenolic ring. The ratio *ortho*/*para* position found was equal to 1.7. An increase in the phenol reactivity was also observed by adding a methylol in the *ortho* position: the *ortho*-hydroxymethylolphenol is twice reactive than the *para*-hydroxymethylolphenol. The dimethylolated compounds were also more reactive, specially the 2,6-dihydroxymethylolphenol.

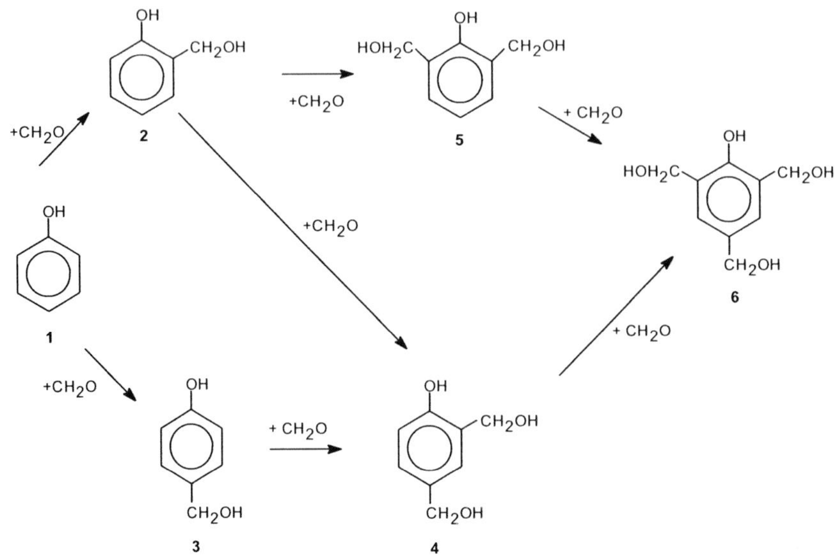

Figure 7: Phenolic methylolated compounds obtained from the reaction between phenol and formaldehyde.

Zsavitsas ([58]; [27]) confirmed the reactivity of the positions determined by Freeman and Lewis ([55]). However, Zsavitsas improved the analysis by gaseous chromatography (GC) of the methylolphenols and incorporated the formaldehyde equilibrium in the kinetic model.

Recently, Grenier-Loustalot and co-workers ([34]) have studied the reaction mechanism of phenol, formaldehyde and model compounds. The reaction of each phenolic methylolated compound with an equimolar quantity of formaldehyde was studied by high-performance liquid chromatography (HPLC) and ^{13}C NMR spectroscopy. There was a competition toward formaldehyde in the addition and condensation reactions between the compounds 2 and 3 (Fig. 7). Most of the dimers obtained were linked by p-p' methylene bridges. The rate of disappearance of the compounds 2 and 3 was equal. The compound 6 was formed from compounds 4 and 5. Then, the condensation reactions occurred with the release of water and/or formaldehyde, forming dimers linked by bridges o-p' and p-p'. Compound 5 was more reactive than compound 4 toward formaldehyde and the difference was attributed to the formation of a chelate:

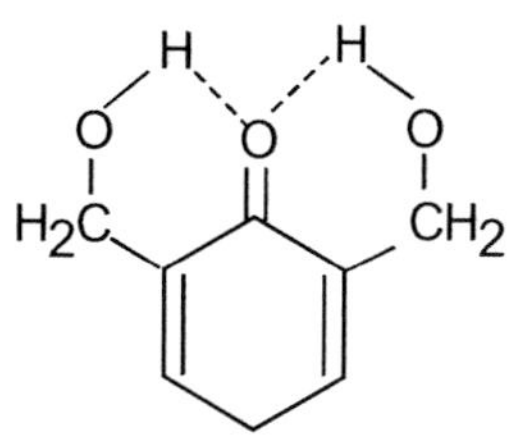

The intramolecular stabilization was due to hydrogen bonding and a higher negative charge is produced in the *para* position.

They determined for the different monomers the following order in the rate constants:

$$k_{phenol} < k_4 < k_2 < k_3 < k_5$$

Compound 5 disappears quickly in reaction medium, whereas the phenol and compound 4 was accumulated in the medium. They found that the *para* position was more reactive than the *ortho* position toward formaldehyde or methylols and led to the formation of methylene bridges o-p' or p-p'; there were no o-o' methylene bridges found. It was improbable that a methylol reacts with a free position of the ring.

By means of computer simulation it is possible to describe the behaviour of the species in the time (e.g.: predict the gelation time). Flory ([62]) represented numerically the phenol-formaldehyde reaction assuming a statistic condensation between the phenol (trifunctional) and the formaldehyde (difunctional). Flory found important differences between the experimental results and the statistics calculus. This is due to simplifications assumed for not consider the different reactivities of the *ortho* and *para* methylol groups and the general change in the reactivitities during the reaction (Knop y Pilato [1]).

Drumm and LeBlanc ([63]) studied the novolak formation but simplified the kinetic scheme. They assumed that there are no present molecules with two or more methylol groups in each chain. They omitted several reactions that really occurred. Kumar et al. ([64]) added this reactions to the model: they considered four types of *ortho* and *para* sites in the phenolic ring (extern: at the chain final, and intern: in the intern ring) and solved equations of mass balance to obtain the phenol and formaldehyde conversions as a function of the time. Then, they determined the molecular weight distribution. These models do not consider the presence of the methylols and as a consequence it is not possible their application to resol resins.

Pal et al. ([65]; [66]) have developed a theoretical model that allows determining the branching reactions of the resols. The mass balance equations were solved numerically to obtain the conversion of the formaldehyde and phenol, number-average chain length, $\overline{M}_w$ and degree of branching as a function of the time. However, the results were not completely valid because the kinetics parameters were taken from the novolak chemistry, which differs widely from the resols in the nature of reactions, reactivity ratios and substitution effects.

More recently, statistic models were developed to predict the distribution of the different species and the statistics parameters in the characterisation of novolaks and resols (Kumar et al. [67]; Williams et al. [68]; Borrajo et al. [69]).

Vázquez et al. ([70]) have developed a statistic model for the formation of the resols using the available kinetic information (Drumm and LeBlanc [16]) and a statistical recurrent method previously used (Aranguren et al. [71]). The model involves nineteen entities (Fig.8) and a mass balance equation for each one has been proposed. These have been solved numerically using the Runge-Kutta technique of the fourth order. They determined the gelation condition, $\overline{M}_n$ and $\overline{M}_w$ and the molecular weight distribution, using NaOH and MgO as catalyst. Different reactivities of the phenolic ring and the substitution effect were taken into account.

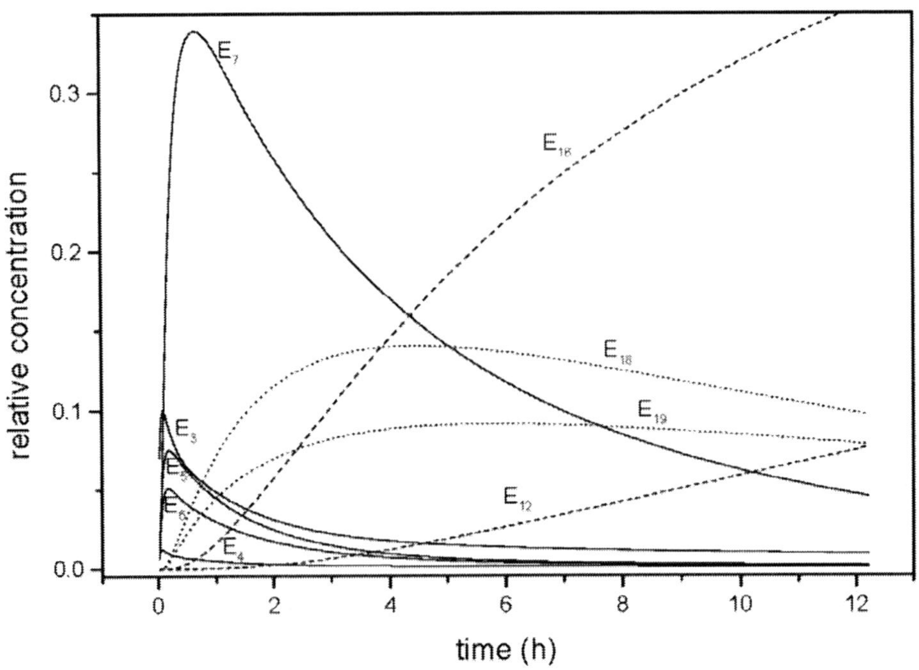

Figure 8: Relative concentration of some entities for a resol having a molar ratio of 1.6. (From Manfredi et al. [73]. Reprinted by permission of Wiley & Sons)

Astarloa-Aierbe et al. ([72]) studied the kinetic of the resol formation catalysed with NaOH, zinc acetate, $Ba(OH)_2$ and triethylamine at a temperature of 80°C, a F/Ph molar ratio of 1.8 and pH=8. They found a good agreement between the experimental results and the ones obtained from the developed kinetic model.

The work began by Vázquez et al. ([70]) have been continued by Manfredi et al. ([73]). They modelled resol-type phenolic resins with different formaldehyde/phenol molar ratio, and compared the results of the model with the chemical structure of the end-products obtained in the polymerisation reaction. Figure 9 shows, as an example, the kinetics of some entities for a formaldehyde/phenol molar ratio of 1.6.

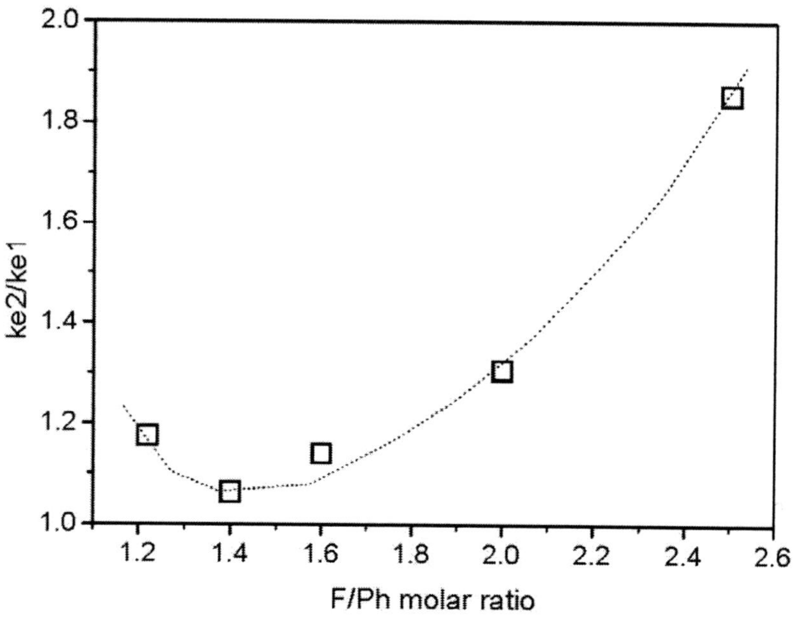

Figure 9: Condensation/addition constant ratio as a function of F/Ph molar ratio. (From Manfredi et al. [73]. Reprinted by permission of Wiley & Sons)

The methylol concentration increases to a maximum at the beginning of the reaction and then decreases (E_4, E_3, E_5, E_6, E_7). It is seen that the different monomer compounds disappear slowly at the expense of dimers. The compounds E_{18}, E_{16} and E_{19} have the highest relative concentrations. The compound with three methylene bridges (E_{12}) increases its relative concentration over the entire reaction period. The experimental results were compared with the ones obtained using the model. A good agreement was found between them. The reactivities ratio determined were similar to the previously reported (Eapen and Yeddanapalli [59]; Freeman and Lewis [56]; Zsavitsas et al. [61]; Zsavitsas and Beaulieu [27]). From the analysis of the ratio of the constants ke_2/ke_1 (that correspond to the phenol-phenolate equilibrium and to the formaldehyde-methylene glycol equilibrium, respectively), it was determined that the condensation reaction occurs in a higher proportion for the resol with a F/Ph molar ratio equal to 1.4 (Fig.9). The activated compound (methylene glycol) is less activated for a resol with a molar ratio of F/Ph: 1.4 indicating that the addition reactions could occur in smaller proportion than the condensation reactions, perhaps because of the higher quantity of the phenol - activated species. As consequence, this resol should present the highest crosslinked structure, as it was found in a previous work (Manfredi et al. [54]). The variation in the relative concentration of the substituted species obtained with the model, was similar to the FTIR results. In addition, the tendency in the polydispersity index showed the same variation during the modification of the F/Ph molar ratio, either measuring by GPC or calculating by the model. The condensation to addition ratio parameter seems to be the most sensitive and important value for fitting the experimental results.

4. Thermal Degradation of Phenolic Resins

Phenolic resins degradation has been fully investigated in the sixties by employing a variety of techniques (Thermogravimetric Analysis, Vapor-Phase Chromatography, X-Ray, Infrared Spectroscopy, Mass Spectroscopy) to examine the reactions involved in the process.

4.1. Mechanism of Degradation Reaction of Phenolic Resins

First, Hall ([74]) reported that the methylene link is the point of attack during oxidation of solid phenol novolaks, by Infrared Spectroscopy studies. Heron ([75]) studied the kinetics of degradation of resols in air by thermogravimetric techniques. Thermogravimetric Analysis (TGA) shows the variation in polymer weight with temperature at a constant temperature rise rate. So, this is an advantageous experiment that provides data over a wide range of temperature with a single test. Also, it shows the different stages of weight loss such as loss of absorbed water, residual monomers and oligomers, water released by cured reactions before the beginning of the polymer degradation (Prime [44]). Heron`s studies indicated that the rate of oxidative degradation of phenolic resins increased with decreasing particle size of the initial sample and with increasing oxygen concentration. He also observed that the rate of weight loss decreased as degradation proceeded, and suggested that this is due to an increase in stability of the resin. The apparent initial activation energy of oxidation was determined and he found that it was of the order of 62.7 kJ/mol (between 300 and 380°C). Between 400 and 600°C the resin ignites and he cannot obtain conclusions about the kinetics of decomposition. Conley and Bieron ([76]; [77]) have studied the oxidative reactions of phenolic resins by infrared spectroscopy at temperatures up to 220°C. They obtained qualitative and quantitative information by analysing the changes in the polymer structure. At this time, a relation between the mechanism of degradation and the resin structure was not established. Also, there was a lack of information about the chemical structure of the cured resin system before the thermal treatment. Then, Jackson and Conley ([78]) investigated the chemistry of a resol resin system at elevated temperatures up to 1000°C. They studied the solid- and gas-phase compositions at the various intervals of temperature by means of Infrared Spectroscopy, Vapor-Phase Chromatography, Thermogravimetric Analysis and X-Ray Diffraction. The infrared spectrum of the phenolic resin before the curing process showed the absence of absorption bands at 6.03, 5.94 and 5.85 μm; these have been assigned to absorption by benzophenone-type carbonyl linkage, quinone-type carbonyl groups and carboxyl carbonyl species respectively (Conley and Bieron [76]). These bands appear from oxidation of the resin, either during its preparation or from exposure to an oxygen-containing atmosphere for extended periods at slightly elevated temperatures. None of these bands were found after maintaining the resin sample during three days at room temperature in air. But after curing it for 3 h at 120°C it was possible to evidence in the infrared spectrum (Fig.10) a rather strong band at 6.03 μm resulting from methylene group ($-CH_2-$) oxidation to benzophenone (C=O).

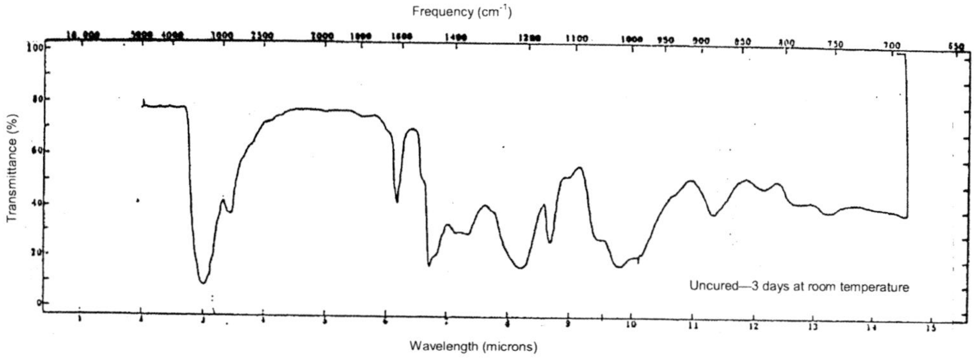

Figure 10: Infrared spectrum of a typical phenolic resin prior and after cure. (From Jackson and Conley [78]. Reprinted by permission of Wiley & Sons)

Further heating of the resin in inert atmospheres at 400°C during 1 h resulted in poscuring and oxidation of the resin. The poscuring was evidenced in the spectrum as a decrease of the band due to the methylol groups (at 9.45 μm) (Fig.11). On the other hand, the additional oxidation was shown as an increase intensity of the band at 5.94 μm due to the quinone-type carbonyl and a broadening of the 3.0-4.0 μm region, together with a weak shoulder on the 5.04 μm band (due to the formation of carboxylic acids).

The products formed on heating the phenolic resin at different temperatures were determined by means of Vapor-Phase Chromatographic studies. They detected the presence of carbon dioxide and carbon monoxide in addition with water, paraformaldehyde, methane and aromatic products as phenol, *ortho* and *para*-cresol, toluene, benzaldehyde and 2,4-dimethylolphenol. Heating the resin up to 400°C, water (result from the condensation of methylol groups) and formaldehyde (result from the loss of methylol groups) were among the products present in major quantities. From these results, Jackson and Conley ([78]) proposed the reaction sequence shown in eq (18 a and b) to afford the presence of these compounds.

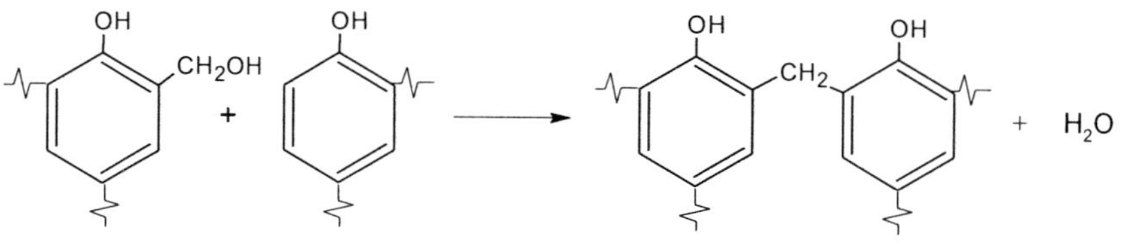

(a)

(18)

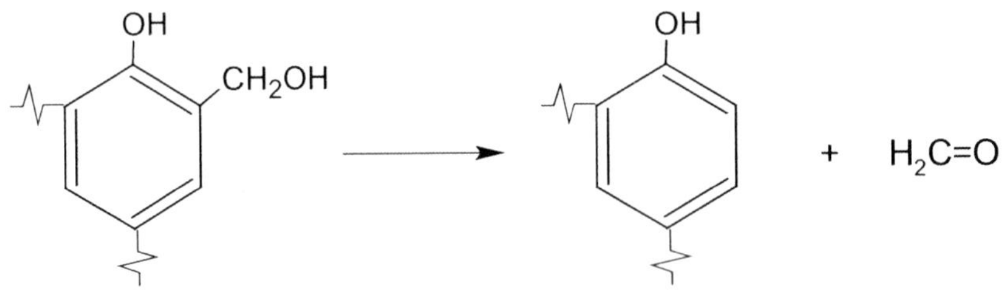

(b)

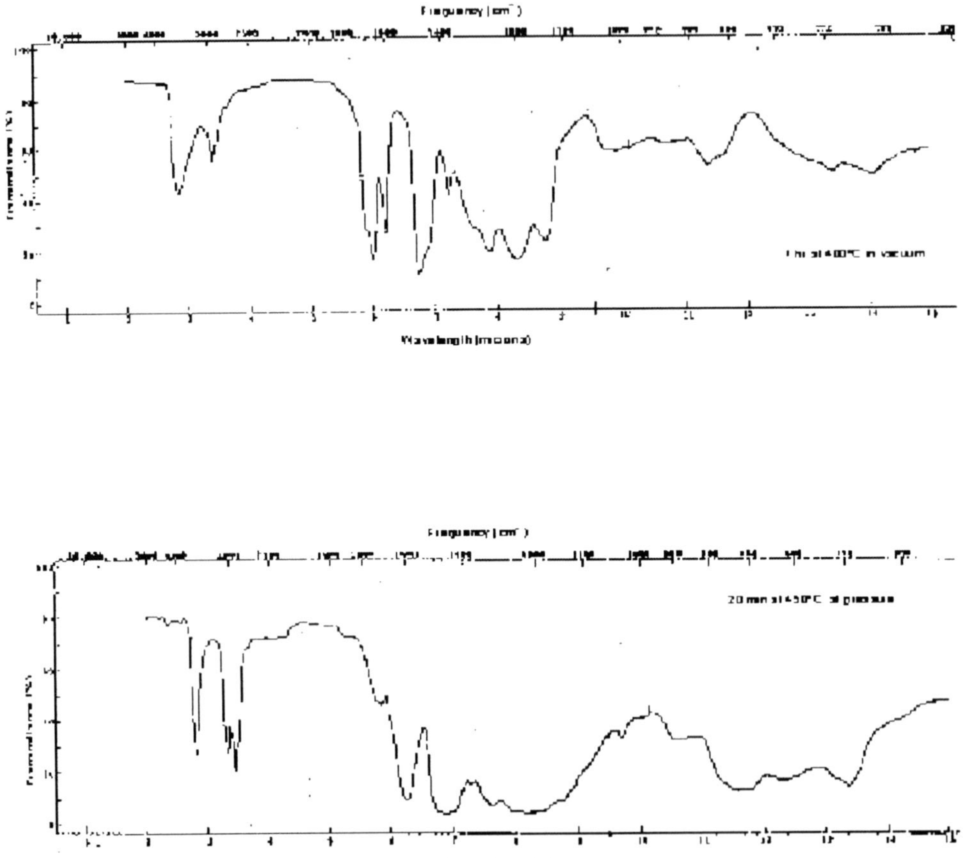

Figure 11: Infrared spectrum of a typical phenolic resin after high temperature exposure. (From Jackson and Conley [78]. Reprinted by permission of Wiley & Sons)

This mechanism is different from the one proposed by Ouchi and Honda ([79]), shown in eq (19), from the study of the pyrolisis of the resin in vacuum.

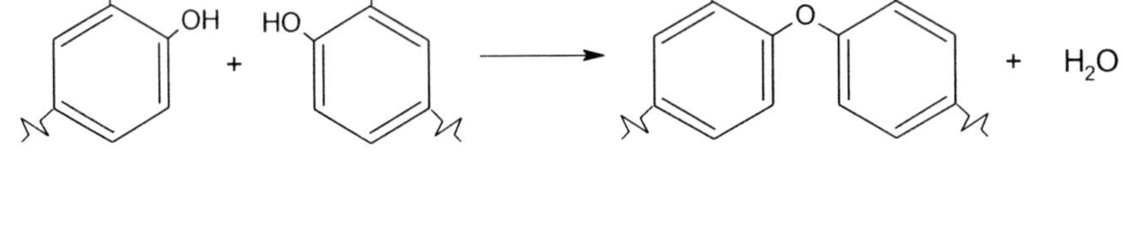

(19)

The mechanism proposed by Jackson and Conley ([78]) is consistent with the decrease in the quantity of methylol groups and the absence of detectable diphenylether linkage in the infrared spectra, in addition with the lower temperature degradation studies made by Conley and Bieron ([76]; [77]).

Analysing the products formed at temperatures higher than 400°C, possible routes for the degradation process were proposed. The predominant reaction route is the oxidation, as evidenced by the production of carbon dioxide and carbon monoxide. The formation of dihydroxybenzophenone linkages and carboxylic acids (as evidenced from the infrared data) from the oxidation of the resin during curing are shown in eq (20) and (21). The initial cured resin contains methylene linkages and appreciable amounts of residual methylol groups:

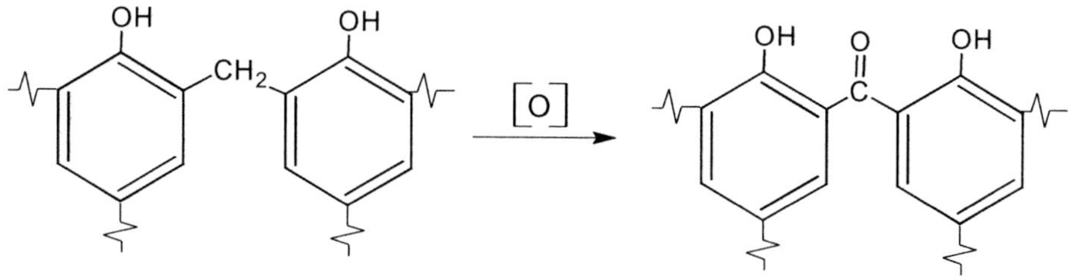

(20)

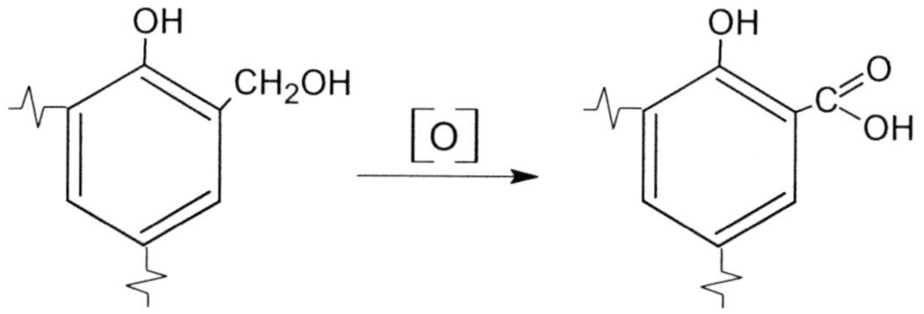

(21)

From these initial oxidation reactions shown in eq (20) and (21), Jackson and Conley ([78]) extended the degradation reaction sequence. Three routes were proposed for the general degradation in a range between 400 and 600°C. Route I, Fig.12, corresponds to the oxidative degradation processes in generalised form (to simplify the diagram there is no indication about the particular position of hydroxyl and methylene group in the aromatic ring). Route II summarises the formation of phenol, cresol, and other methyl-substituted species, together with methane. Methane is formed in considerable amounts above 400°C, whereas cresol and xylenol type species are generally found in relatively constant amounts, as the temperature increases. Route III summarises the formation of benzene, toluene, and benzaldehyde, obtained by the loss of hydroxyl from the aromatic ring. The hydroxyl radical formed, could act as a source of oxygen for further oxidation reactions as well as a source of water reacting with a hydrogen from any available methylene linkage. It was found that the oxidative mechanism is faster than the fragmentation reactions up to 1000°C.

Decomposition to produce char and carbon monoxide simultaneously, via ring scission, occurs rapidly at temperatures above 600°C. Figure 13 shows the char-forming reaction by the decomposition of the oxidised resin through a quinone-type intermediate. Char formation was detected visually and by the presence of a graphite-like line in the X-Ray pattern of the residue. Also, in the charred material at 700°C, the presence of hydroxyl and carbonyl groups was observed.

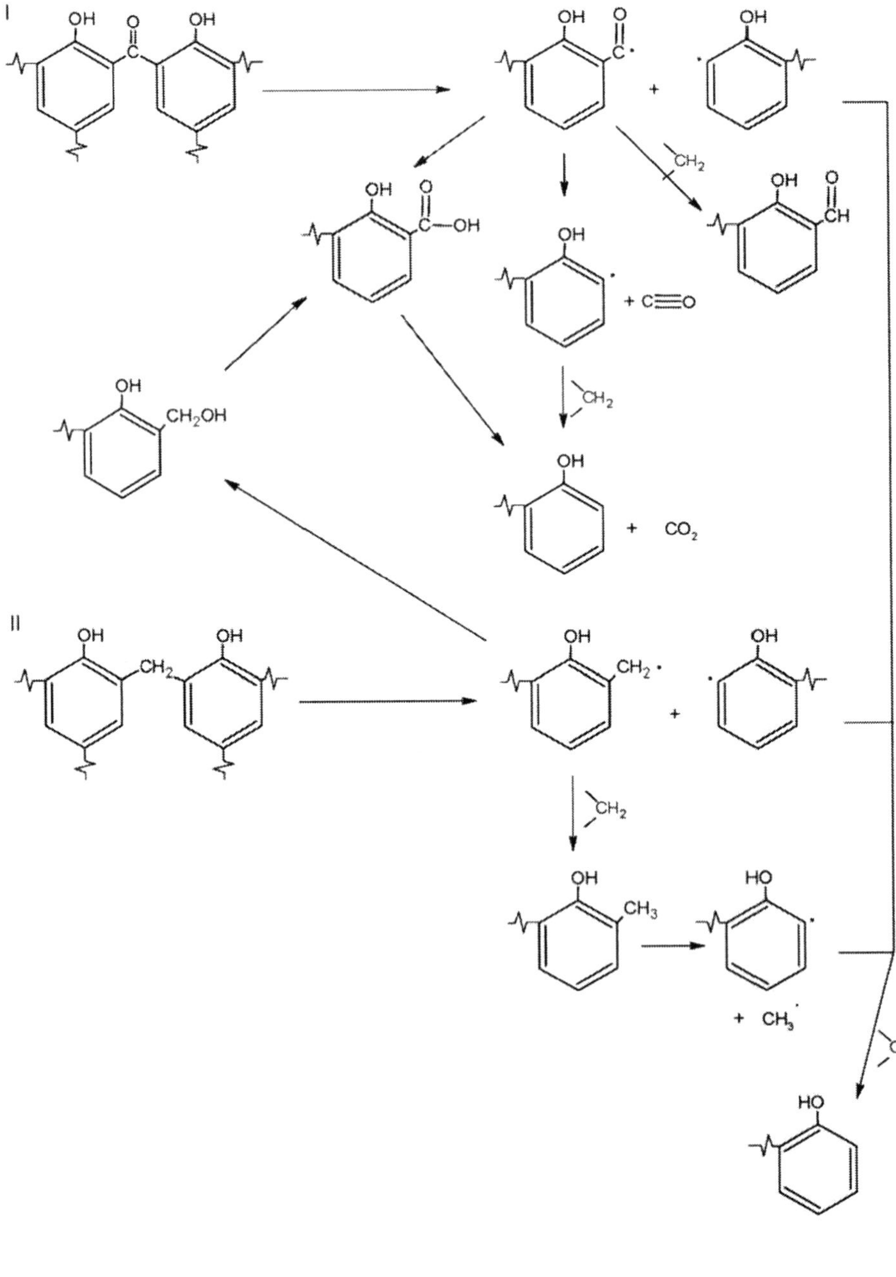

Figure 12: Proposed reactions for the decomposition at high temperatures. Route I: oxidative degradation processes; Route II: fragmentation reactions; Route 3: formation of the derivatives of the benzene. (From Jackson and Conley [78]. Reprinted by permission of Wiley & Sons)

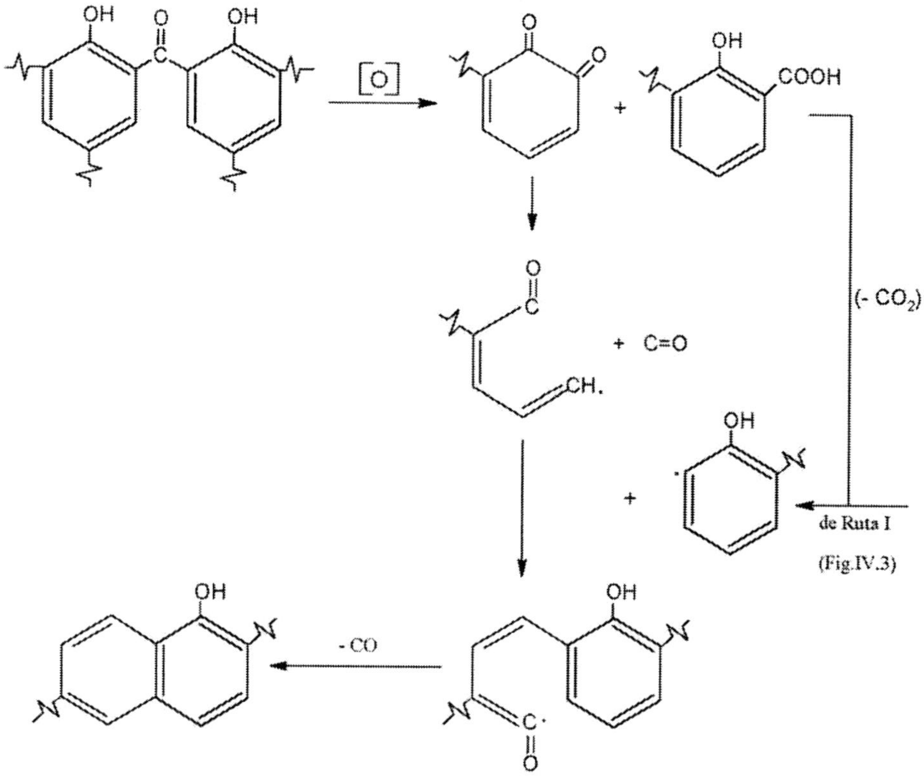

Figure 13: Char-forming reaction. (From Jackson and Conley [78]. Reprinted by permission of Wiley & Sons)

Lotche et al. ([80]) corroborate the char forming mechanism previously proposed. They investigated the degraded residue composition as a function of temperature by combining the weight loss information with elemental analysis. Resols with a F/Ph molar ratio between 2 and 2.7 and in the temperature range between 150 and 800°C, were studied. They found that regardless of the species present in the liquid resol before curing, once fully cured, all of them degrade in the same way. The degradation processes occurring at elevated temperatures seems to be associated with the stability of the dihydroxydiphenylmethane unit. Only at temperatures up to 300°C small differences in the degradation mode were observed due to the poscuring processes and volatilisation of solvent and other low molecular weight species.

Recently, Lenghaus et al. ([81]) proposed a new mechanism of ether breakdown to explain the formation of methyl and carbonyl groups. They showed that neither of the mechanism previously described in the literature (Maciel et al. [30]) could be adequately applied to their ^{13}C CP-MAS NMR results.

The previously mechanism involved the oxidation of an hydroxymethyl group to an aldehyde:

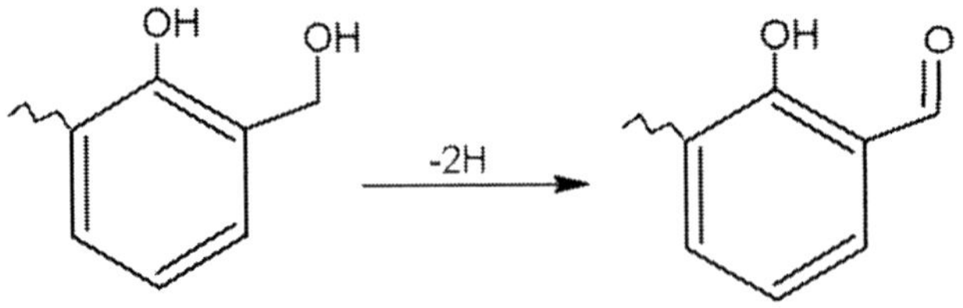

(22)

or the scission of an ether bridge to generate an aldehyde and a methyl group:

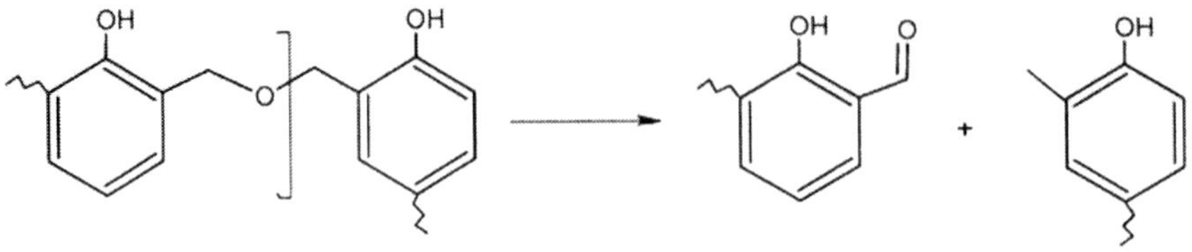

(23)

The new mechanism involves an ether scission generating a hydroxymethyl phenol and quinone methide, which subsequently reacts to give a carbonyl and a methyl group, respectively.

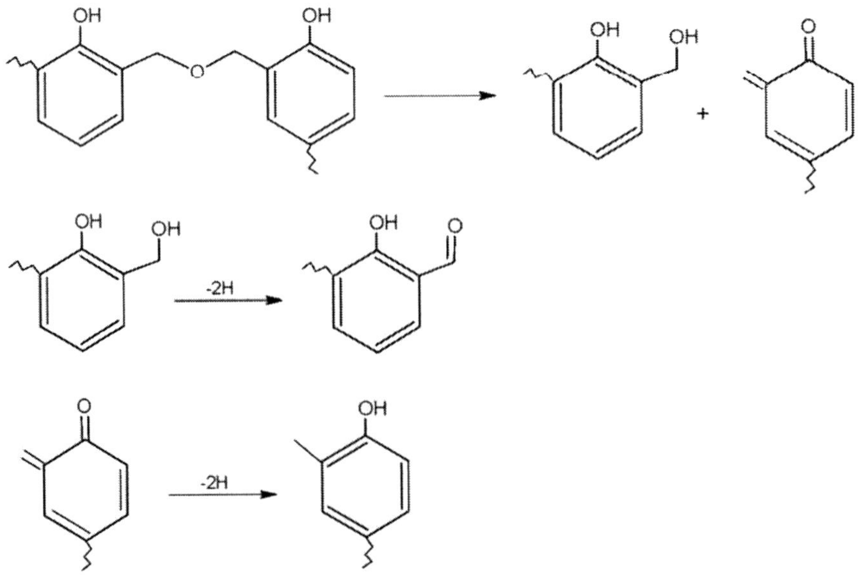

(24)

They compared two resols with different F/Ph molar ratio: 1.2 and 1.8. Only in the resol with F/Ph=1.2 cured up to 205°C there was no presence of carbonyl, phenoxy or methyl groups by ^{13}C CP-MAS solid state NMR. This was interpreted due to larger quantities of ether bridges present in the F/Ph=1.8 than in the F/Ph=1.2. They studied the surface area and microscope structure of the two resols after carbonisation up to 1000°C. They found that the resol with F/Ph=1.2 has a narrower microscope structure and size distribution than the resol with F/Ph=1.8.

Shulman et al. ([82]) using mass spectrometric thermal analysis to 800°C studied the composition and kinetics of formation of the volatile products formed from the degradation of resol type phenolic resins. Activation energies for formation of each product and for general degradation of the resins were determined. Activation energies for formation of each product were calculated from an Arrhenius plot (linear semilogarithmic graph of rate of formation versus reciprocal temperature). Those who showed higher activation energy were the phenolic homologs products of thermal scission of methylene-phenyl bonds. Oxidation products showed to have activation energies similar to or slightly higher than that of oxidation of methylene bridges to carbonyl groups. Activation energies obtained were 46-67 kJ/mol for formaldehyde, 71 kJ/mol for water and 79 kJ/mol for CO_2. The high activation energies for formation of phenol, cresol and xylenol (313.5±105 kJ/mol) and the molar ratios in which they are formed are the evidence that phenolic products are formed by a thermal fragmentation process. It was determined that the general degradation of phenolic resins is a first-order reaction. From the slope of an Arrhenius plot, the activation energy was determined to be of the order of 92 kJ/mol. Other activation energy values have been previously reported by Friedman ([83]) (230 kJ/mol) and by Farmer ([84]) (54-84 kJ/mol). The difference in the values was attributed to different cure cycles.

More recently, in a series of publications, Chiantore et al. ([85]; [86]; [87]) studied the thermal decomposition of phenol-formaldehyde resins using different techniques as Fourier transform infrared spectroscopy (FTIR), chromatography and thermogravimetric analysis. They have analysed ([85]) the carbonaceous residue obtained by pyrolisis of phenolic samples at temperatures up to 600°C in an inert atmosphere by using diffuse reflectance Fourier transform infrared spectroscopy. Analysing the residue formed, new functional groups were present together with some characteristic groups of the starting material.

The reactions of branching and fragmentation proposed for the pyrolisis are shown in eq (25) and (26)

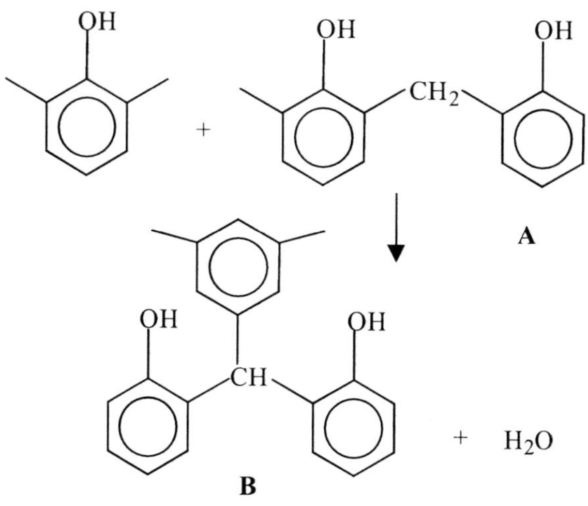

(25)

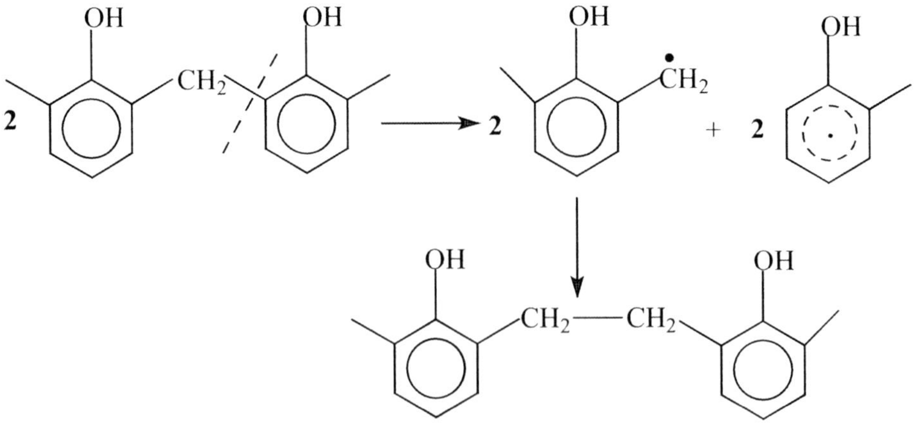

C

(26)

On the spectra of the residue (Fig.15) the bands of the phenolic structures appeared: a broad band near 3540 cm^{-1} corresponds to the OH stretching and the narrower band at 1320 cm^{-1} and the relative strong one at 1220 cm^{-1} are indicative of the OH deformation and CO stretching respectively. The CH stretching of the aromatic structures appears as a band above 3000 cm^{-1}; and the bands at 820 and 760 cm^{-1} correspond to the aromatic out-of-plane deformations and vibrations. The presence of the structures identified as A, B and C in eq (25) and eq (26) is justified by the band near 2900 cm^{-1} in the spectra, that corresponds to the CH stretching of the aliphatic structures; and the deformation of the CH_n structures is attributed to the band near 1470 cm^{-1}.

Jackson and Conley ([78]) have proposed the formation of benzophenone units as shown in eq (20). These ketonic structures should appear in the spectra as two bands near 1700 and 1660 cm^{-1}, but they appear only as shoulders on the side of the strong peak near 1600 cm^{-1}. This suggests that this oxidative reaction does not occur frequently during pyrolisis. The strong peak at 1600 cm^{-1} corresponds to the C=C stretching in a polyaromatic system. Due to the band observed at 1260 cm^{-1}, Chiantore et al. ([85]) proposed the presence of diphenyl ether structures as shown in eq (19), and as Ouchi and Honda ([79]) did.

In another work, Chiantore and Guaita ([86]) combined the thermal and chromatographic techniques to determine the novolak resins composition. They analysed the products obtained after heating the sample at 150, 270 and 500°C under vacuum by Size Exclusion Chromatography (SEC). As the temperature increases, the molecular weight of the fractions evolved from the resin became higher because the higher the chain fragmentation and the easier evaporation. The chromatogram showed the same position of the peaks than the original sample. It agrees with previous reports (Camino et al. [88]) based on Infrared Spectroscopy analysis. This results show that the chemical structure of chain fragments after a thermal treatment of a novolak resin is similar to the original structure of the resin.

More recently, Chiantore et al. ([87]) investigated the thermal behaviour of an uncured and a cured novolak type resin under different experimental conditions by TGA. They found that the weight loss of an uncured resin occurs in several stages under nitrogen flow (Fig.14). The volatilisation of residual monomers and oligomers occur below 200°C in the same manner that with the resol resins. Then, there is a competition between decomposition and crosslinking reactions at temperatures up to 1000°C and formed the charred structure. Three temperatures (320, 410 and 550°C) of maximum weight loss, partially superimposed can be seen. On the other hand, when the analysis was made in air flow, a high peak appeared at 630 °C and there was a complete sample fragmentation into volatile products.

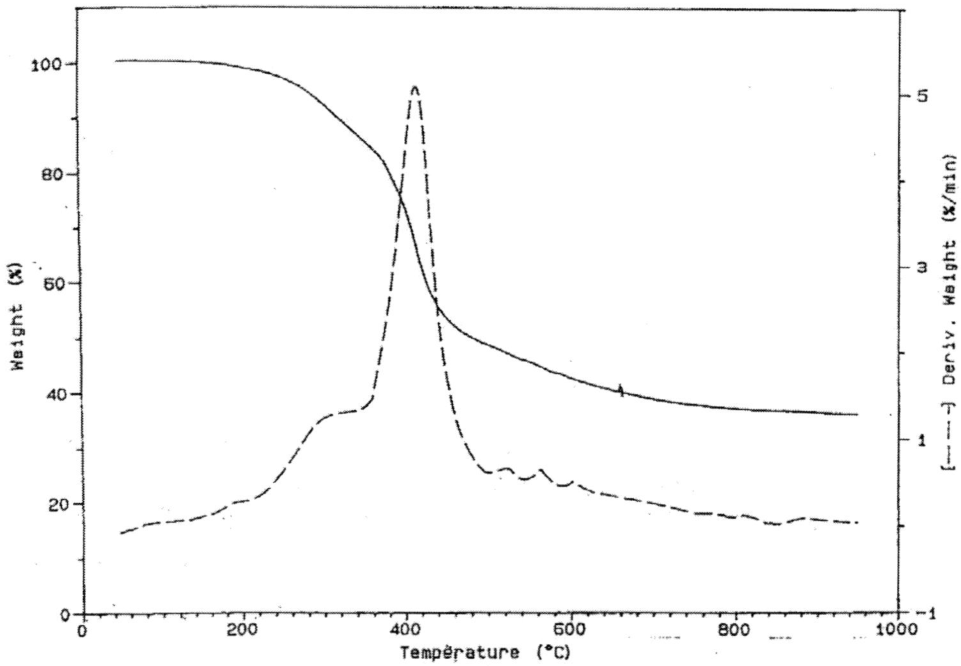

Figure 14: TGA and DTG curves of novolak under nitrogen flow. (From Chiantore et al. [87]. Reprinted by permission of Gordon and Breach Publishing)

Then, the novolak resins were cured with HMTA at 250°C and their thermal behaviour was analysed by TGA. Under nitrogen flow the decomposition occurs in two principal stages, at 420 and 550°C (Figure 15). These two stages could be related to random homolytic scission followed by fragmentation and crosslinking reactions; and to further decomposition of structures previously formed, leading to formation of a charred aromatic residue. The degradation process involving free-radical reactions together with volatilisation of higher-molecular weight oligomers occurs at temperatures higher than 400°C. By comparing Figures 14 and 15 it can be notice that the curves pattern are similar but the residual weight at 1000°C is higher for the cured novolak due to their higher crosslinking density.

Under air, the resin behaviour is similar to that obtained under nitrogen flow, at temperatures up to 350°C. In this case, at higher temperatures, the weight loss occurs in three main stages: at 420, 557 and 633°C. At 700°C there was a complete decomposition to volatile products.

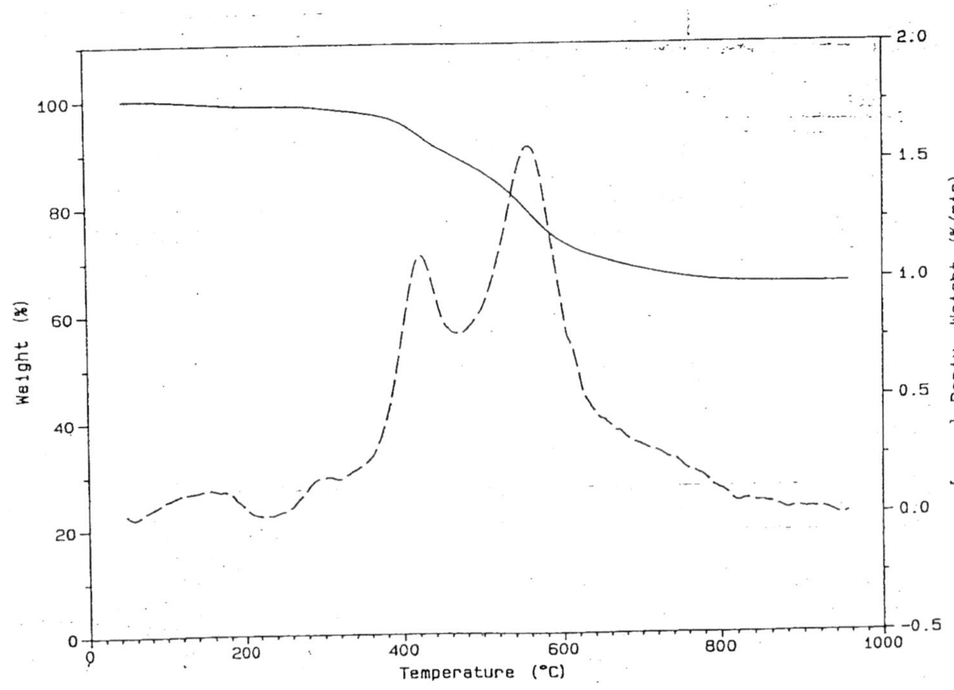

Figure 15: TGA and DTG curves of cured novolak under nitrogen flow. (From Chiantore et al. [87]. Reprinted by permission of Gordon and Breach Publishing)

The principal volatile products obtained under nitrogen and air were: carbon dioxide, formaldehyde, benzene, toluene, m-xylene, phenol, o-cresol, p-cresol and xylenols, together with water, methane and carbon monoxide. Among these products, toluene and p-cresol were the main ones formed under nitrogen flow, and it was benzene in air flow. The nature of the products is the same either in air or in an inert atmosphere but the relative amounts depend on the degree of crosslinking and on atmospheric conditions.

More recently, Puglia et al. ([89]) studied the relationship between chemical structure of resol resins and their thermal and fire resistance. Thermal degradation of a serie of resol type phenolic resins with a wide range of formaldehyde/phenol molar ratio (F/Ph): 1.0, 1.2, 1.3, 1.4, 1.6, 2.0 and 2.5 was studied. They were thermally cured up to 190°C. Thermogravimetric analyses were done under nitrogen flow from room temperature to 900°C. Typical weight changes in the cured materials in function of the temperature are shown in Figure 16. The residual mass obtained agrees with the literature values (Shulman and Lochte [82]).

Figure 17 shows the residual mass of each resol in function of the F/Ph molar ratio. The maximum residual mass was obtained for the F/Ph molar ratio equal to 1.4. In a previous work (Manfredi et al. [54]), obtained a maximum quantity of methylene bridges for a resol with a similar F/Ph molar ratio, related with the stoichiometric ratio between the phenol and formaldehyde. As a consequence, in the condensation reaction, the phenol acts with a functionality less than 3.0. Commercial resol has been studied to compare their behaviour with the synthesised ones. They showed the same residual mass than the synthesised ones with the same molar ratio. As a consequence, the concentration of catalyst, differences on reaction time or temperature are not so important as the initial molar ratio, within certain range.

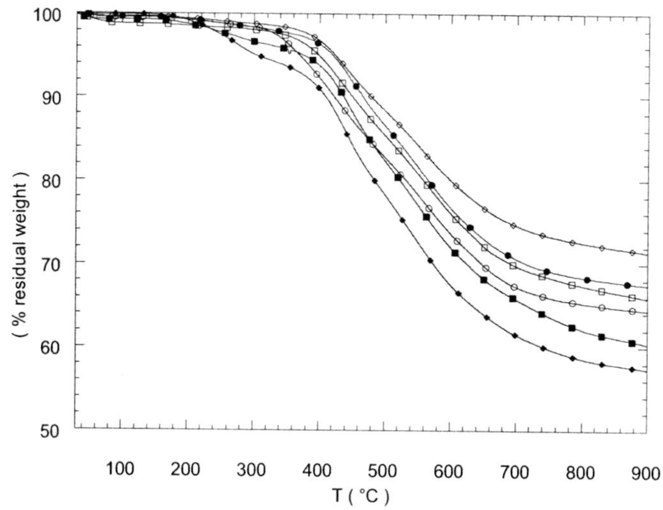

Figure 16: Thermogravimetric curves at 10°C/min of resol with different molar ratio F/Ph: (O) 1.2, (□) 1.3, (□) 1.4, (●) 1.6, (■) 2.0 and (□) 2.5.

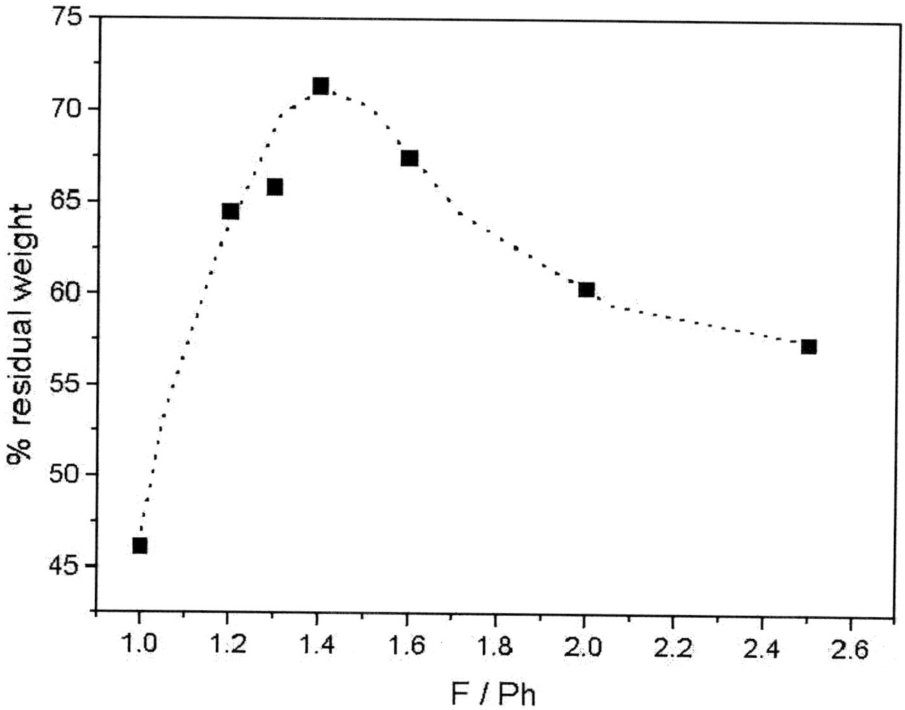

Figure 17: Residual mass of each resol in function of the molar ratio F/Ph.

The derivated results of the mass loss in function of temperature for each resol were plotted in the Figure 18.

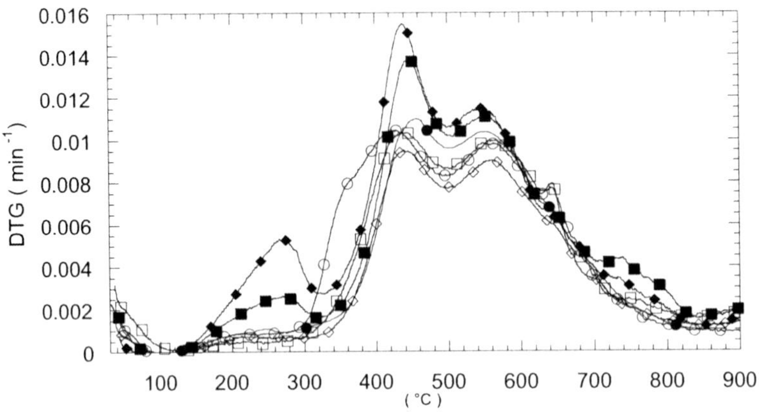

Figure 18: DTG curves at 10°C/min of resol with different molar ratio F/Ph: (O) 1.2, (□) 1.3, (□) 1.4, (●) 1.6, (■) 2.0 and (□) 2.5.

It is possible to distinguish three regions. Weight losses below 300°C correspond to volatilisation of entrapped solvent, water from methylol condensation, and low molecular weight component (Lochte et al. [80]). From 350°C to 650°C there are two peaks. The height of the peak at aprox.430°C is changing with the molar ratio of the resol in the series. The height of the peak at aprox. 550°C is quite the same for all the resol. These two peaks of the serie of resols were separated to analyse them. Table 1 shows the height and the residual mass of the three peaks for the different resols. The height of the peak that appears below 300°C is minimum for F/Ph=1.4. This peak could be attributed to formaldehyde, phenol and oligomers and may be to pendant chain scission. The height of the peak at aprox. 430°C has a minimum value for F/Ph=1.4 and the residual mass has a maximum value for the same resol. This peak could be related with the methylene bridges content, because the resol with a F/Ph molar ratio close to 1.4 should have the highest content of bridges. The height and mass loss of the peak at aprox. 550°C are quite constant. This peak should be related with other degradation reaction, because it does not seem to depend on the methylene bridge content. It could be more related with the ArOH reaction proposed by Chiantore et al. ([85]).

FTIR analysis was performed on the residual mass of three resols heated during 1 h at 400 °C in a TGA instrument and the results are reported in the Figure 19. A previous study of a novolak phenolic resin was taken as reference (Morterra and Low [90]) for the assignment of the IR peaks. Figure 19 also shows the FTIR of the residual mass of the resol with F/Ph 2.0 after poscuring it during 1 h at 600 °C. The presence of the same bands and a decrease in their height should be noticed. The decrease of the bands between 1400 and 1500 cm^{-1} seems to be related with the scission of the CH-CH structures. The presence of the A, B and C compounds shown in eq (29 and 30) proposed by Chiantore et al. ([85]) could be possible to identify. The presence of the different types of compounds at the two temperatures studied indicates that it is not possible to separate the reactions of degradation and identify each compound with the temperature.

Table 1: Separation of the Three Peaks of the Studied Resol in TGA at 10°C/Min

F/Ph	dα/dt peak below 300°C	α_{final}%	dα/dt peak at aprox. 430°C	α_{final} %	dα/dt peak at aprox. 550°C	α_{final} %
1.0	0.00150		0.00076		0.00096	92.
1.2	0.00007	99.36	0.00097	85.1	0.00064	92.1
1.3	0.00004	98.93	0.00094	88.3	0.00070	91.9
1.4	0.00003	99.36	0.00065	94.0	0.00070	86.3
1.6	0.00010	98.35	0.00090	91.8	0.00070	94.0
2.0	0.00018	97.66	0.00100	91.1	0.00069	91.2
2.5	0.00049	93.34	0.001250	88.3	0.00075	91.6

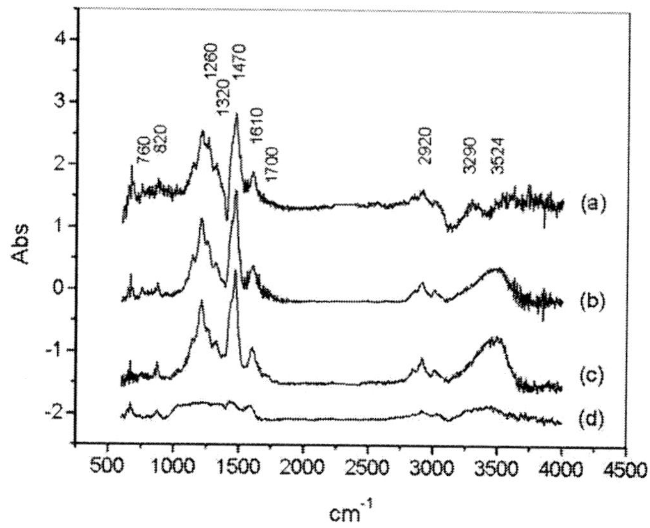

Figure 19: FTIR spectra for resol at 400°C with molar ratio F/Ph: (a) 1.0, (b) 1.3 and (c) 2.0, and at 600°C: (d) 2.0

4.2. Kinetics and Lifetime Predictions

Absolute classification of thermal stability is difficult because of the myriad of chemical and physical processes that can complicate the overall kinetics to be measured (Flynn and Dickens [91]; Flynn [92]). Decomposition mechanisms, especially oxidative degradation are diffusion controlled, so that sample geometry and fillers can alter the decomposition process. Rather than being an absolute measure, this type of information should be used as a guide for further studies of the decomposition behaviour in the use of phenolic resins as matrix in composite or coatings.

Lifetime predictions require an experimentally determined relationship between time to failure and appropriate variables that force the failure (e.g. temperature and environmental factors). The most common accelerating variable is temperature, which is related to time by the activation energy. Flynn ([93]) studied the application of the thermal analysis methods to lifetime predictions of polymeric systems (Prime [44]).

Up this time, kinetic degradation of resol resin with a wide range of formaldehyde to phenol molar ratio was not widely studied. So, Puglia et al. ([89]) applied different methods developed in the literature to determine the kinetics parameters from thermogravimetric analysis data. Firstly, considered the degradation process by a simple n order kinetic expression:

$$\frac{d\alpha}{dt} = ko\exp(-\frac{E_a}{RT})(1-\alpha)^n \text{ n=1,2,.......} \tag{27}$$

where;

$$\alpha = \frac{w_o - w(t)}{w_o - w_f}$$

wo is the initial mass, *w(t)* is the mass at a certain time, and w_f is the final mass.

The equation (27) is similar to the following:

$$\frac{d\alpha}{dT}\frac{dT}{dt} = ko\exp(-\frac{E_a}{RT})(1-\alpha)^n \tag{28}$$

where dT/dt is called β (heating rate).

A model with order n=1 was proposed for resol resin degradation (Shulman and Lochte [82]). As consequence, the following equation was proposed:

$$\ln\frac{d\alpha}{dt(1-\alpha)} = \ln ko - \frac{E_a}{RT} \tag{29}$$

This equation was applied to determine the activation energy of the two peaks that appear in the 350-650°C range of the TGA analysis for the serie of resols with F/Ph from 1.2 to 2.5 (Figure 18). Figure 20 shows the graph of equation (33) for the peaks at aprox. 430°C and 550°C corresponding to resol with F/Ph molar ratio 1.2. The activation energies were obtained from the slopes of the negative parts. The same procedure was performed for each resol. The values of the activation energies computed are reported as a function of the F/Ph molar ratio in Figure 21. The activation energy obtained for both peaks have a maximum on a F/Ph=1.6. As a consequence, both peaks are related with the methylene bridge content (Manfredi et al. [54]).

Other methods that were developed to determine the kinetic parameters from the thermogravimetric analysis data (Kissinger [94], Flynn and Wall [95], [96], Friedman [83]) are not so simple as the previously used. The isoconversionals multiple heating rate methods allow estimation of the activation energy from relationships between the heating rate and the temperatures to reach constant conversion. These methods could be applied for systems whose activation energy is independent of temperature but may change with conversion. These are useful to yield kinetic information on systems with multiple reactions, as in this case: overlap of degradation with curing process.

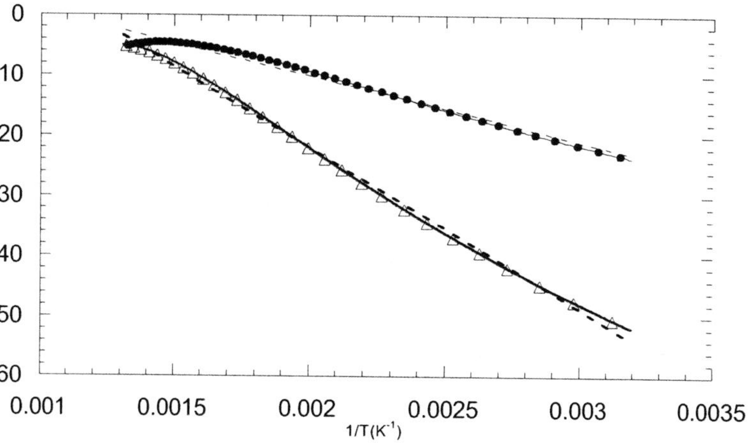

Figure 20: Plot of ln(d□/dt / (1-□)) vs 1/T for a resol with a molar ratio of 1.2: (●) peak at aprox. 430°C, (□) peak at aprox. 550°C.

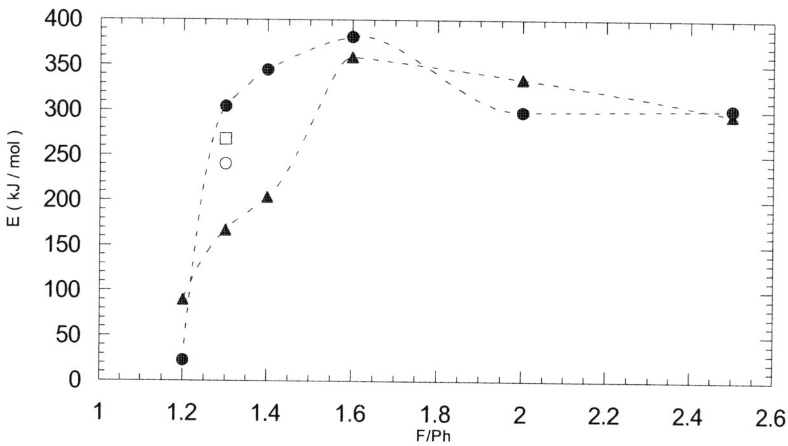

Figure 21: Values of activation energy for the: (▲) peak at aprox. 430°C, and (●) peak at aprox. 550°C (model with n=1), (O) peak at aprox. 430°C, (□) peak at aprox. 550°C (Kissinger model).

All of these methods start with eq (34):

$$\frac{d\alpha}{dt} = f(\alpha) ko \exp(-\frac{E_a}{RT}) \tag{30}$$

$$\frac{d\alpha}{dt} = \frac{A}{\beta} \exp(-\frac{E_a}{RT}) f(\alpha) \tag{31}$$

Kissinger method ([94]) uses the temperature at which the rate of weight loss is maximum, from which the kinetic parameters are obtained. It leads to the following expression:

$$\frac{d(\ln \beta / T_m^2)}{d(1/T)} = -\frac{E_a}{R} \tag{32}$$

where T_m is the maximum peak temperature. Thermograms of the resol with a F/Ph molar ratio of 1.3 obtained at several constant heating rates were normalised (Fig.22) and the activation energies for the two peaks in the range 350-650°C were obtained applying the Kissinger method previously described. The correlation between the heating rate and the maximum of temperature was plotted in Figure 23. The activation energies values obtained by this method, which does not require any assumption on the value of the reaction order, agree with the values previously obtained with the simple model considering a reaction order n=1

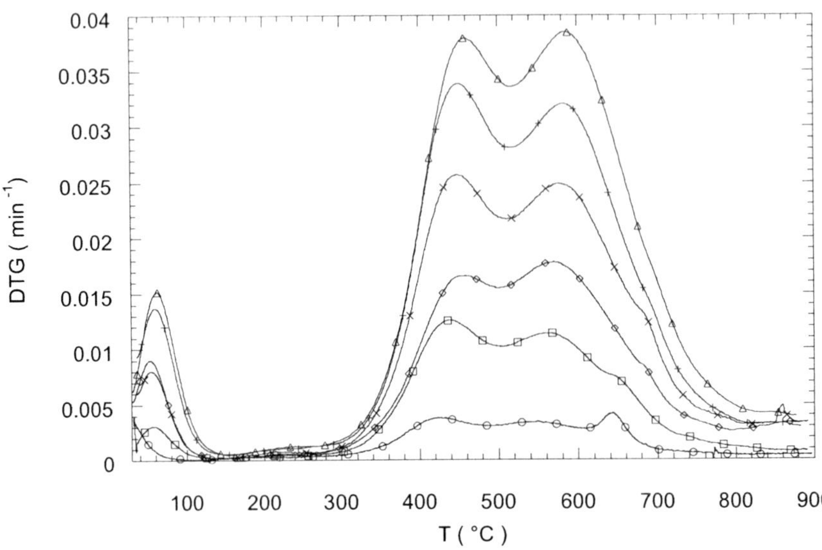

Figure 22: DTG curves of resol with F/Ph=1.3 at different heating rates: (O) 5°C/min, (□) 10°C/min, (□) 15°C/min, (X) 20°C/min, (+) 25°C/min and (□) 30°C/min.

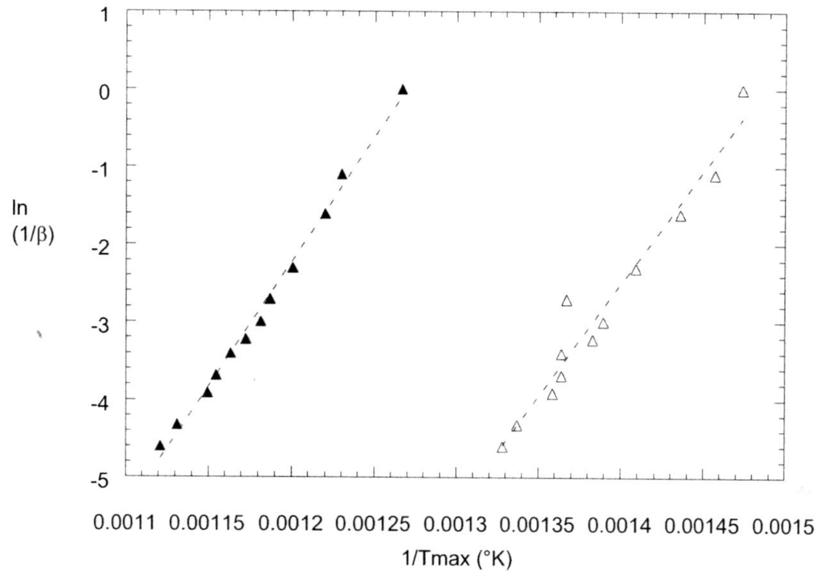

Figure 23: Application of Kissinger method to the degradation of the resol with a F/Ph molar ratio=1.3 for the: (▲) second peak, and (□) third peak.

The Flynn and Wall method ([95], [96]) also provides activation energies without any assumption about the reaction order. This method assumes a generic equation, eq (35), which after some mathematic manipulation leads to the following operative expression:

$$\left(\frac{d \ln \beta}{d \frac{1}{T}} \right)_{\alpha} = -1.052 \frac{E_a}{R} \tag{33}$$

The apparent activation energy E_a can be obtained from plotting lnβ versus 1/T for a constant α, since the slope of such line is given by -1.052 Ea/R.

Table 2 shows the activation energies obtained by applying the Flynn and Wall method on the two peaks in the 350-650°C range of the thermograms shown in Figure 22. The activation energy changes with the conversion, a higher conversion leads to higher activation energy for the degradation process. This behaviour is consistent with the fact that when the temperature increases, the degree of crosslinking also increases and a more stable polymer is produced.

Table 2: Activation Energies by the Flynn and Wall Method for the Resol with a F/Ph Molar Ratio of 1.3 and Different Heating Rates

α	Ea_1 (kcal/mol) (peak at aprox.430°C)	α	Ea_2 (kcal/mol) (peak at aprox.550°C)
0.1	120.1	0.43	190.2
0.13	126.9	0.5	200.0
0.23	145.6	0.53	216.1
0.30	157.2	0.6	225.2
0.33	164.4	0.7	249.4
0.40	178.0	0.73	268.0
0.43	190.2	0.8	297.7

The Friedman method ([83]) was also applied to estimate the apparent kinetic parameters from the TGA data of the resol with F/Ph 1.3 (Figure 22). Friedman provided the following estimation:

$$\ln\left(\frac{d\alpha}{dt}\right) = \ln A + \ln f(\alpha) - \frac{E_a}{RT} \tag{34}$$

The apparent activation energy at a given degree of conversion can be determined from the slope of the plot on the left hand side of eq (34) versus reciprocal temperature for different heating rates. Figure 24 shows the values of the activation energies obtained by the Friedman and Flynn and Wall methods for the resol with F/Ph 1.3. The results obtained using both methods were very similar.

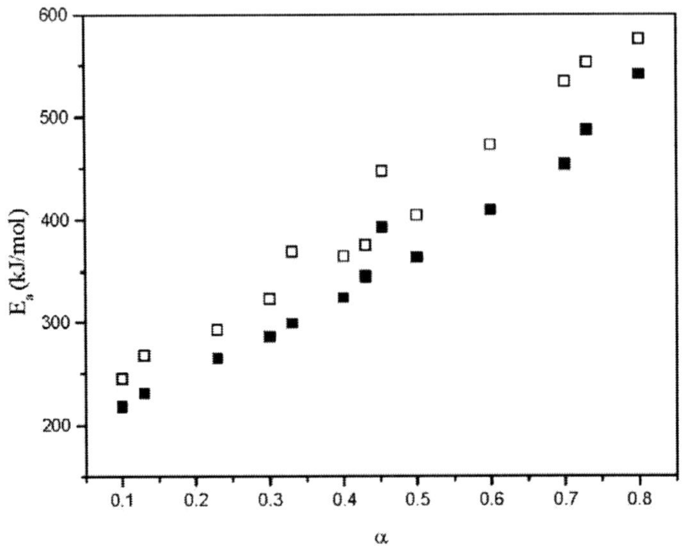

Figure 24: Values of activation energy obtained by: (□) Friedman method and (■) Flynn and Wall method.

4.3. Fire Resistance, Smoke Developed and Flame Retardants of Phenolic Resins

Combustion is a chemical phenomenon in which an oxidant is reacted with a fuel to produce energy and numerous combustion products. Uncontrolled or accidental combustion is called fire. The course of the fire can be split into several phases (Troitzsch [97]): the start of fire, the initiation of fire and the fully developed fire (Figure 25). The initiating fire is produced by an ignition source. This material burns and generates heat, heating up and igniting additional combustible matter. At this stage radiant heat and the temperature have increased to such an extent that the material decomposes and produce the fire spread over the entire area (flash over). The flash over leads to the fully developed fire, which constitutes the second phase. In the second phase, the temperature reaches the 1000°C and the fire has reached its peak. Depending on the air, the fire passes its climax and enters the decreasing phase (Kracklauder [98]).

Smoke generation is one of the hazards associated with material's flammability, and this has recently received only relatively attention. The term smoke has a vague definition and several connotations. In general use, smoke is considered to be a cloud of particles, individually visible, which opaque due to the scattering and/or absorption of visible light. Another definition of smoke is the composite of gaseous products of burning material in which small solid or liquid particles are also dispersed, it relates the phenomenon to combustion and introduces the aspect of gaseous products. However, it is necessary to distinguish between combustion gases and visible smoke since the two have different sources and effects, as well as, different methods of measurement and significance in fire scenarios.

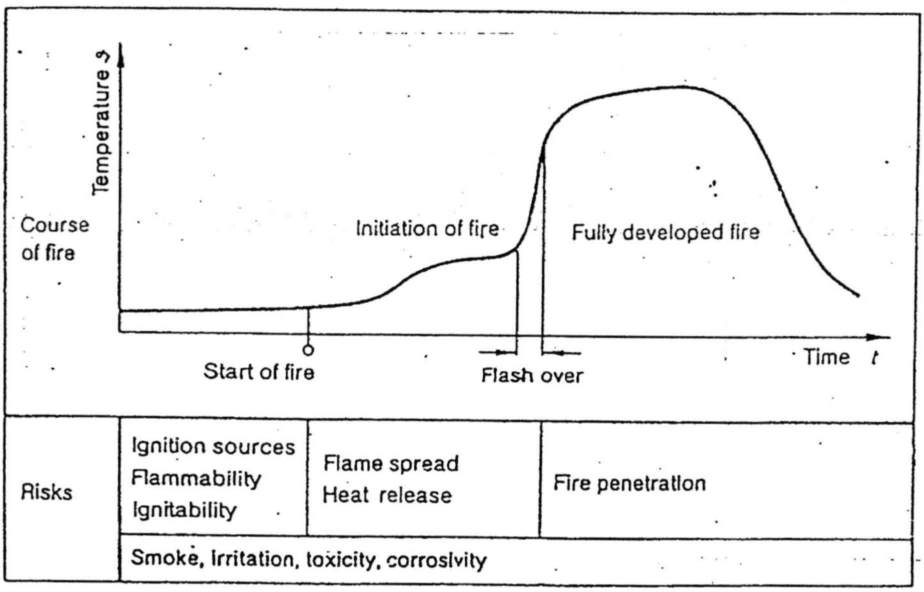

Figure 25: Phases involved in the course of a fire.

Among the combustion gases, carbon monoxide, CO, has commonly been of chief concern, with several reports examining its role in detail. Other toxic gases that can be formed in fires include hydrogen cyanide, nitrogen oxides, hydrogen chloride, and sulfur oxides (Lawson [99]).

Polymers with high thermal stability or which form limited amounts of flammable pyrolysates generally have limited visible smoke. Char formation is one conceivable mechanism for minimising the yield of pyrolysates.

The development of smoke and toxic gases is so-called secondary fire effects that occur in conjunction with the combustion process. The level of smoke development depends on numerous factors such as the source of ignition, oxygen availability, the constitution and properties of the combustible material. Smoke posses a danger to human life, particularly in the initial phases of a fire, due to the loss of orientation caused by sight obscuration. This can lead to panic and irrational behaviour which makes escape impossible with fatal results. Death rarely occurs as consequence of the inhalation of smoke particles, but is generally due to the inhalation of toxic gases, lack of oxygen and heat radiation (Troitzsch [100]). Pyrolisis and combustion of thermosets results in minimal smoke development. Phenolic resins have a lesser tendency to emit smoke since they char intensely and only small amounts of aromatic are liberated in the gas phase. As it was explained before they decompose by statistical chain rupture yielding mainly formaldehyde, phenol an other oxygen-containing fragments such as acetone and alcohols. Phenolic resins are normally employed as ablators, for high temperature thermal protection systems. So, an understanding of the degradation processes, specially those reactions leading to char formation, is very useful.

The fire resistance of a series of resols was analysed by Puglia et al. ([89]) determining the mass loss in a Cone Calorimeter for three different F/Ph molar ratios: 1.0, 1.3 and 2.0 as reported in Figure 26.

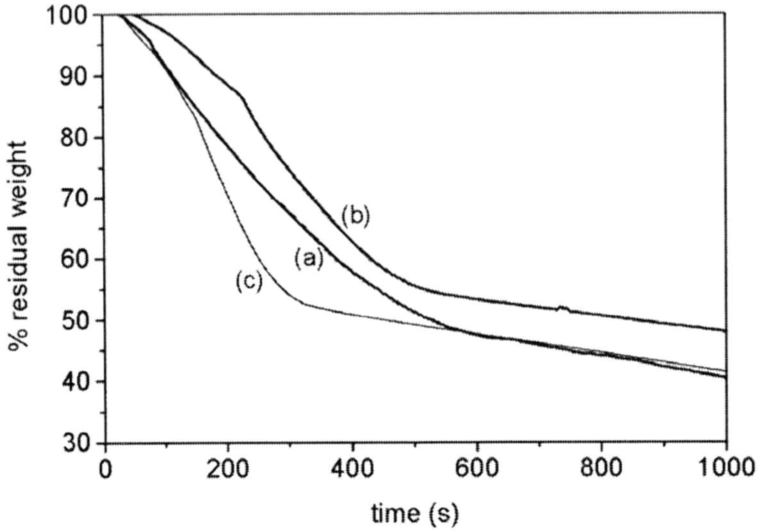

Figure 26: Residual mass on cone-calorimetry for resols with different F/Ph molar ratio: (a) 1.0, (b) 1.3, and (c) 2.0.

ASTM E 1354-90 was taken into account for the test measurements. It is evident that initially, the degradation process is controlled by the chemical reaction. After certain time the mechanism of degradation is affected by mass transfer and changes from chemical control to diffusion control (Gann et al. [101]). This fact can be attributed to the formation of a char in the skin of the specimen, reducing the thermal conductivity and promoting the diffusion control of the degradation process. The results of the Cone Calorimeter tests are reported in Table 3 in terms of residual mass at two different temperatures and in function of the F/Ph molar ratio. Before 400 s, the residual mass value is maximum for the more crosslinked resol, F/Ph 1.3. The lower value corresponds to the resol with F/Ph molar ratio 2.0. However, after 900 s, the resol with F/Ph=1.0 has the minimum value. This means that the resol with F/Ph=2.0 may have the lower conductivity, due to the higher quantity of pendant chain or methylol groups and the higher foaming process growing from the volatile formed during the firing process which leads to higher char quantity.

Table 3: Loss Mass of the Cone Calorimeter Equipment for Different F/Ph Molar Ratio Resol Resin

F/Ph	% residual mass
	t=400s
1.0	58
1.3	63
2.0	51
	t=900s
1.0	42
1.3	49
2.0	43

The heat release rate (RHR) for different resol resins was determined (Figure 27) and an increase with the formaldehyde content was obtained (Table 4). Most of the heat was released by the resols after the first ignition point or flash point. However, the resol with a F/Ph molar ratio of 2.0 has a different behaviour than the others. Immediately after ignition heat was developed. The time between the first ignition and the main propagation could be important in order to stop the fire. Table 4 shows the results of ignition time ($t_{ignition}$) and the time in the maximum of the peak of propagation (t_{max}). The time of ignition is related with the initiation of fire and the t_{max} can be related with the time of the propagation of the fire. The ignition time was higher for the resol with F/Ph=1.3 than for the other resols, due to their highest crosslinking density (Manfredi et al. [54]).

The area under the RHR curve is the energy associated to fire development. The area of the sample is broken during the firing process due to the volatilisation of the low molecular weight compounds, as a consequence this value is not so precise (Table 5). However, the highest value was obtained for the resol with molar ratio F/Ph = 2.0.

Another useful result of this evaluation is the velocity of smoke production (SEA), Table 6. Also in this case, the resol with F/Ph=1.3 shows the best behaviour. The evolution of the gas was also determined (Table 7). The resol with a F/Ph molar ratio 1.3 has the lower CO_2 evolution and not present evolution of CO. Resol with the higher F/Ph molar ratio presented the higher quantity of CO_2 and CO evolved.

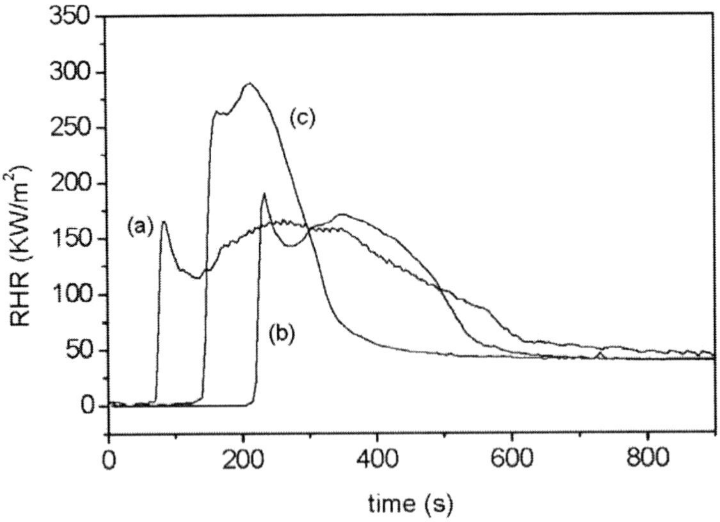

Figure 27: Heat release rate for resols with different molar ratio F/Ph: (a) 1.0, (b) 1.3, and (c) 2.0.

Table 4: Results of the Time of Ignition and Time in the Peak of Propagation and RHR for Different Resols

F/Ph	t_{max} (s)	$t_{ignition}$ (s)	RHR ignition (kW/m^2)
1.0	262	70	166.5
1.3	349	222	190.3
2.0	215	148	260

Table 5: Release Heat of Each Resol

F/Ph	Total heat release (kJ)	Total heat release/initial mass (kJ /gi)
1.0	600.5	25.3
1.3	469.8	24.0
2.0	588.4	28.5

Table 6: Rate of Smoke Production for the Resols (SEA)

F/Ph	Peak SEA	Peak time (s)	$SEA_{average}$
1.0	0.471	346	0.231
1.3	0.4211	460	0.215
2.0	0.7882	300	0.294

Table 7: Results of the Average and Peak of Emission of CO_2 y CO

F/Ph	CO_2 peak (kg/kg)	CO_2 average (kg/kg)	CO peak (kg/kg)	CO average (kg/kg)
1.0	0.62	0.20	0	0
1.3	0.39	0.18	0	0
2.0	1.13	0.45	1.32	1.32

A flame-retardant should inhibit or even suppress the combustion process. Depending on their nature, flame retardants can act chemically and/or physically in the solid, liquid or gas phase. They interfere with combustion during a particular stage of this process, e.g. during heating, decomposition, ignition or flame spread (Troitzsch [102]). An ideal fire-retardant system will: a) minimise adverse effects on desirable properties, such as strength, colour, stability, processability, etc, b) have versatility across a product line and show good durability in service and c) cause no increase in cost, preferably lowering costs. The existence of any such ideal system is uncertain, but it represents a goal for development work (Lawson [99])

In the few cases where phenolic resins require fire retardant treatment, additive or reactive flame-retardants have proved effective. Besides familiar flame-retardants such as tetrabromobisphenol A and various organic phosphorus compounds used in many plastics, halogenated phenols and aldehydes (e.g. p-bromobenzaldehyde) are specifically utilised as reactive flame-retardants. Phosphorous can be introduced by direct reaction of the phenolic resin with phosphorus oxychloride. Likewise inorganic compounds such as boric acid are incorporated into phenolic resin by chemical reaction (Troitzsch [102]). In order to suppress the afterglow of phenolic resins, compounds like aluminium chloride, antimony trioxide and organic amides are used.

4.4. Thermal Degradation of Phenol-Formaldehyde Resins from Renewable Resources

Vegetable tannins are natural products of relatively high molecular weight, which have the ability to complex strongly with carbohydrates and proteins. They are the most important natural products used industrially, specifically in leather tannin processes and in the synthesis of wood adhesives to replace phenol in phenol-formaldehyde adhesives (Pizzi [103]). Tannins obtained from chestnut bark, wattle (mimosa), quebracho extracts and cashew-nuts have been used (Kulvik [104], Roux [105]). The simple phenols available in these tannins are mainly pyrogallol, ellagic acid, gallic and m-digallic acids (Figure 30). The distribution of the substituent group on the benzene nuclei of these phenols ensures that their reactivity toward formaldehyde is lower than of synthetic phenol. It is then true that hydrolysable tannin-formaldehyde resins can be prepared in the same traditional manner as the synthetic phenolic resins of commerce, but their preparation time will be much longer. At comparable reaction times, the distribution of the condensates formed is shifted toward lower molecular weights, negatively affecting the cured strength of the resin, even when methyloled, their reactivity will be lower. The pre-polymer obtained by means of the prereaction route of phenol and formaldehyde is not possible in the case of tannin resins. The high reactivity of the flavonoid A-ring of mimosa and most of the other polyflavonoid structure, ensure the condensation reaction. The methylol groups condense with any reactive species present. If only tannins are present, they condense quickly with other flavonoid A-rings to form methylene bridges.

O'Connor and Blumm ([106]), studied the effect of the side chain on the thermal stability of phenolic resins. Cashew liquids and other substituted phenols were copolymerised with phenol and formaldehyde under acid catalyst to yield novolak-type prepolymers. The substituted phenols used were o-cresol, p-cresol, p-nonylphenol, and m-penta-decylphenol. The resins were subsequently cured with hexamethylenetetramine to give polymers. They determined by thermogravimetric analysis that the higher the amount of phenol in the resin, the higher its thermal stability. The main degradation of the raw cashew resins occurs between 450°C and 500°C. The thermal stability of a series of substituted phenol compounds with different length of the alkyl side chain was investigated. The obtained results showed that the thermal stability decreases when the alkyl chain length increases. So, the order of thermal stability of resins increases in the order: pentadecylphenol < nonylphenol < cresols < phenol. The level of hexamethylenetetramine used to cure the polymer also influences the stability of the resin. A level of 10-15% curing agent maximise the thermal stability of the network.

Barks of calabrian pine (Pinus brutia) and Anatolia chestnut (Cestanea sativa) tree species have been used to replace formaldehyde in the synthesis of novolak resins. These species were condensed with phenol in the presence of sulfuric acid as catalyst at a temperature of 130°C during 1h. The phenolate bark species obtained were cured with hexamethylenetetramine at 190°C during 5 min and then, thermogravimetric and calorimetric analysis were performed on them. The sulfuric acid concentration was varied from 0.5 to 4.0%. The TGA results showed that the weight loss of the cured phenolated barks decreased with increasing catalyst concentration. In the same way, there was a decrease in the glass transition temperature of cured barks with increasing sulfuric acid concentration. The thermogravimetric weight loss of the cured phenolated barks was found to be comparable to those of cured commercial novolak resin and phenolated wood (Alma and Kelley [107]).

ACKNOWLEDGEMENTS

The authors gratefully acknowledge CONICET and Fundación Antorchas of Argentina for financial support.

REFERENCES

[1] Knop A, Pilato LA. In: *Phenolic Resins*. Springer-Verlag Berlin Heidelberg, 1985
[2] Wu HD, Ma C-CM, Chu PP. *Polymer* 1997; 38(21):5419-5429
[3] Yang TP, Kwei TK, Pearce EM. *J.Appl.Polym.Sci.* 1991; 41:1327
[4] Matsumoto A, Hasegawa K, Fukuda A, Otsuki K. *J.Appl.Polym.Sci.* 1991; 43:365
[5] Matsumoto A, Hasegawa K, Fukuda A, Otsuki K. *J.Appl.Polym.Sci.* 1992; 44:205
[6] Matsumoto A, Hasegawa K, Fukuda A, Otsuki K. *J.Appl.Polym.Sci.* 1992; 44:1547
[7] Wu HD, Ma C-CM, Lee MS, Wu YD. *Die Angew Makromol. Chemie.* 1996; 35:235
[8] Tyberg CS, Bergeron K, Sankarapandian M, Shih P, Loos AC, Dillard DA, McGrath JE, Riffle JS, Sorathia U. *Polymer* 2000; 41:5053-5062

[9] Gardziella A, Pilato LA and Knop A. In: *Phenolic Resins*, Springer-Verlag Berlin Heidelberg, 2000
[10] Wilder RV. *Revista de Plásticos Modernos* 1988; 389:771
[11] Brown N, Linnert E, Buhl D. *Composites* 1996; 14:29
[12] St John NA, Brown JR. *Composites Part A* 1998; 29^a:939
[13] Martin RW. In: *The Chemistry of Phenolic Resins*. New York: J. Wiley, 1956
[14] Knop A, Scheib W. In: *Chemistry and Applications of Phenolic Resins*. Springer-Verlag Berlin Heidelberg, 1979
[15] Aranguren MI, Borrajo J, Williams RJJ. *J.Polym.Sci., Polym.Chem.Ed.* 1982; 20:311-318
[16] Drumm MF, LeBlanc JR. *Kinetic Mech.Polym.* 1972; 3:157
[17] Shipp DA, Solomon DH. *Polymer* 1997; 38(16):4229
[18] Moragues J. In: *Analyse, Etude Cinetique et Trasformations Structurales de Resols.* PhD Tesis, INSA, Lyon, France, 1994
[19] Kopf PW, Little AD. In: Mark H, Bikales N, Overberger C, Menges G, Kroschwitz J, editors. *Encyclopedia of Polymer Science and Engineering*. New York: Wiley & Sons, 1985. 2nd Edition, Vol 11
[20] DeJong JI, DeJonge J. *Rec. Trav. Chim.* 1953; 72:497
[21] Von Euler H, Dekispezy SV. *Z.Phys.Chem.A.* 1941, 189:109
[22] Jones TT. *J.Soc.Chem.Ind.* 1964; 69:102
[23] Bassow NJ et al. *Plaste and Katschuk* 1974; 6:417
[24] Bender H. *Ind.Eng.Chem.* 1952; 44:1619
[25] Higginbottom HP, Culbertson HM, Woodbrey JC. *J.Polym.Sci.*, Part A. 1965; 3:1079
[26] Hatfield GR, Maciel GE. *Macromolecules* 1987; 20:608
[27] Zsavitsas A, Beaulieu A. *Am.Chem.Soc.Div.Org.Coat.Plast.Chem.Pap.* 1967; 27:100
[28] Peer HG. *Rec.Trav.Chim.* 1959; 78:851
[29] Peer HG. *Rec.Trav.Chim.* 1960; 79:825
[30] Maciel GE, Chuang I-Ssuer, Gollob L. *Macromolecules* 1984; 17:1081-1087
[31] So S, Rudin A. *J.Appl.Polym.Sci.* 1990; 41:205
[32] Grenier-Loustalot MF, Larroque S, Grenier P, Leca JP, Bedel D. *Polymer* 1994; 35(14):3046-3054
[33] Grenier-Loustalot MF, Larroque S, Grande D, Grenier P, Bedel D. *Polymer* 1996; 37(8):1363-1369
[34] Grenier-Loustalot MF, Larroque S, Grenier P, Bedel D. *Polymer* 1996; 37(6):939-953
[35] Grenier-Loustalot MF, Larroque S, Grenier P, Bedel D. *Polymer* 1996; 37(6):955-964
[36] Astarloa-Aierbe G, Echeverría JM, Egiburu JL, Ormaetxea M, Mondragón I. *Polymer* 1998; 39(14):3147-3153
[37] Astarloa-Aierbe G, Echeverría JM, Martín MD, Mondragón I. *Polymer* 1998; 39(15):3467-3472
[38] Astarloa-Aierbe G, Echeverría JM, Vázquez A, Mondragón I. *Polymer* 2000; 41:3311-3315
[39] Gupta MK, Hindersinn RR. *Polym.Eng.Sci.* 1987; 27(13):976
[40] Kumpinsky E. *Ind.Eng.Chem.Res.* 1995; 34(9):3096-3101
[41] Kim MC, Amos LW, Barnes EE. *Polymer* 1990; 24:173

[42] Holopainen T, Alvila L, Rainio J, Pakkanen TT. *J.Appl.Polym.Sci.* 1997; 66:1183-1193

[43] Rudin A, Fyfe C, Martin Vines S. *J.Appl.Polym.Sci.* 1983, 28:2611-2622

[44] Prime B. Thermosets. In: Turi, editor. *Thermal Characterization of Polymeric Materials*, Academic Press, 1997. 2nd edition, chapter 6, p. 1710-1711

[45] Benson M, Jura L. *Org.Magn.Reson.* 1984; 22(2):86

[46] Myers GE, Christiansen AW, Geimer RL, Follensbee RA, Koutsky JA. *J.Appl.Polym.Sci.* 1991; 43:237

[47] Kim MG, Nieh WL-S, Meacham RM. *Ind.Eng.Chem.Res.* 1991; 30:798-803

[48] Follensbee RA, Koutsky JA, Cristiansen AW, Myers GE, Geimer RL. *J.Appl.Polym.Sci.* 1993; 47:1481-1495

[49] Christiansen A W, Gollob L. *J.Appl.Polym.Sci.* 1985;30:2279

[50] Wang X-M, Riedl B, Christiansen A W, Geimer R L. *Polymer* 1994;35(26):5685-5708

[51] Focke WW, Smit MS, Tolmay AT, Van der Walt, LS, Van Wyk, WL. *Polym.Eng.Sci.* 1991; 31(23):1665-1669

[52] Kenny JM, Pisaniello G, Farina F, Puzziello S. *Thermochimica Acta* 1995; 269/270:201-211

[53] Latocha C, Balcerowiak W. *J.Appl.Polym.Sci.* 1990; 40:13

[54] Manfredi LB, de la Osa O, Galego Fernández N, Vázquez A. *Polymer* 1999; 40(13): 3867-3875

[55] Jones TT. *J.Soc.Chem.Ind.(London)* 1946; 65:264

[56] Freeman J, Lewis CW. *J.Am.Chem.Soc.* 1954; 76:2080

[57] Yeddanapalli LM, Kuriakose AK. *J.Sci.Ind.Res.(Ind)* 1959; 18B:467

[58] Zsavitsas AA. *Am.Chem.Soc.Div.Org.Coat.Plast.Prep.* 1966; 26:93

[59] Eapen KC, Yeddanapalli LM. *Makromol.Chem.* 1968; 119:4

[60] Zsavitsas AA. *J.Polym.Sci.A-1* 1968; 6:2533-2542

[61] Zsavitsas AA, Beaulieu, RD, LeBlanc JR. *J.Polym.Sci.A-1.* 1968; 6:2541-2557

[62] Flory PJ. *Chem.Rev.* 1946; 39:137

[63] Drumm MF, LeBlanc JR. In: Solomon DH, editor. *Step Growth Polymerization.* New York: Marcel Dekker, 1972

[64] Kumar A, Kulshreshtha AK, Gupta SK. *Polymer* 1980; 21:317

[65] Pal PK, Kumar A, Gupta SK. *Polymer* 1980; 22:1699

[66] Pal PK, Kumar A, Gupta SK. *The British Polymer Journal* 1981; 12:121

[67] Kumar A, Gupta SK, Kumar B. *Polymer* 1982; 23:1929

[68] Williams RJJ, Adabbo HE, Aranguren MI, Borrajo J, Vázquez A. *Polym.Prep. (Am.Chem.Soc., Div.Polym.Chem.)* 1983; 24:169-170

[69] Borrajo J, Aranguren MI, Williams RJJ. *Polymer* 1982; 23:263

[70] Vázquez A, Adabbo HE, Williams RJJ. *Ind.Eng.Chem.Prod.Res.and Dev.* 1984; 23: 375-379

[71] Aranguren MI, Borrajo J and Williams, RJJ. *Ind.Eng.Chem.Prod.Res.Dev.* 1984; 23:370

[72] Astarloa-Aierbe G, Mondragón I, Riccardi C. *Simposio Argentino de Polímeros 1999*; Los Cocos, Córdoba, Argentina. P175

[73] Manfredi LB, Riccardi C, de la Osa O and Vázquez A. *Polymer International 2001*; 50:1-7

[74] Hall RW. *Chem.& Ind.* 1952; :693
[75] Heron F. Pyrolisis of a Phenol Formaldehyde Polycondensate, SCI Monograph N° 13. In: *Thermal degradation of Polymers*, Society of Chemical Industry : 475-499
[76] Conley RT, Bieron JF. *J.Appl.Polym.Sci.* 1963; 7:103
[77] Conley RT, Bieron JF. J.*Appl.Polym.Sci. 1963*; 7:171
[78] Jackson WM, Conley R.T., *J.Appl. Polym.Sci. 1964*; 8:2163-2193
[79] Ouchi K, Honda H. *Fuel 1959*; 38:429
[80] Lochte HW, Strauss EL, Conley RT. *J.Appl.Polym.Sci.*, 1965; 9:2799-2810
[81] Lenghaus K, Quiao GG, Solomon DH. *Polymer 2001*; 42:3355-3362
[82] Shulman GP, Lochte HW. *J.Appl.Polym.Sci.* 1966; 10:619-635
[83] Friedman HL. *J.Polym.Sci.* Part.C 1964; 6:183
[84] Farmer RW. *Technical Documentary Report n° ASD-TDR-62-1043*, PartI, 1963; p.32
[85] Chiantore O, Novák I, Berek D. *Anal.Chem.* 1988; 60:638-642
[86] Chiantore O, Guaita M. *J.Appl.Polym.Sci.: Appl.Polym.Symp.* 1991; 48:431-440
[87] Chiantore O, Lazzari M, Fontana M. *Int. J. Polymer Analysis and Characterization (IJPAC),* 1995; 1(2):119-130
[88] Camino G, Luda MP, Costa L, Trossarelli L In: *Proc. 7th Int.Conf. Thermal Analysis.* Miller B, ed. New York: J.Wiley, 1982. Vol 2, p.1137
[89] Puglia D, Kenny JM, Manfredi LB, Vázquez A. *Materials Engineering 2001*; 12(1):55-72
[90] Morterra C, Low MJD. *Carbon 1985*; 23 : 525-530
[91] Flynn JH, Dickens B. *Thermochim.Acta* 1976; 15:1-16
[92] Flynn JH. *J.Therm.Anal.* 1988; 34:367-381
[93] Flynn JH. *J.Therm.Anal.* 1995; 44:499-512
[94] Kissinger H.E., *Analyt.Chem.*, 1957; 29:1702-1706
[95] Flynn JH, Wall LA. *J.Polym.Sci.*, Part.B 1966; 4:323
[96] Flynn JH, Wall LA. *J.Res.Natl.Bur.Stand.*, Sect.A 1966; 70:487
[97] Troitzsch J. The Burning Process. In: *International Plastics Flammability Handbook*, New York: Hanser Publishers, 1990. 2nd edition. Chapter 3, p. 11-15
[98] Kracklauder J. Smoke and Tenability: A Perspective on the Materials Appoach to the Fire Problem. In: Lewin M, Atlas SM, Pearce M, editors. *Flame-Retardant Polymeric Materials.* New York: Plenum Press, 1978. Vol 2, Chapter 9
[99] Lawson DF. Methods for Reduction of Smoke from Burning Polymers. In: Lewin M, Atlas SM and Pearce M, editors. *Flame Retardant Polymeric Materials.* New York: Plenum Press, 1982. Vol.3, Chapter 2, p.39-41
[100] Troitzsch J. Smoke Development. In: *International Plastics Flammability Handbook.* Hanser Publishers, 1990. 2nd edition. Chapter 6, p. 63-74
[101] Gann R, Dipert R, Drews M, In: Mark H, Bikales N, Overberger C, Menges G, Kroschwitz J, editors. *Encyclopedia of Polymer Science and Engineering*, New York: Wiley & Sons, 1985. 2nd edition, Vol 7.
[102] Troitzsch J. Fire Retardant Plastics. In: *International Plastics Flammability Handbook.* Hanser Publishers, 1990. 2nd edition. Chapter 5, p. 43-52
[103] Pizzi A. In: Hemingway RW and Larks PE, editors. *Plant Polyphenols, Synthesis, Properties and Significance.* New York and London: Plenum Press, 1992. p.991-1003
[104] Kulvik E, *Adhesives Age* 1977; 20(3):33-34

[105] Roux DG. Reflections on the Chemistry and Affinities of The Major Commercial Condensed Tannins in the Context of Their Industrial Use. In: Hemingway RW, Larks PE, editors. *Plant Polyphenols, Synthesis, Properties and Significance*. New York and London: Plenum Press, 1992. p. 7

[106] O'Connor D, Blum FD. *J.Appl.Polym.Sci.* 1987; 33:1933-1941

[107] Alma MH, Kelley SS. *Polymer Degradation and Stability 2000*; 68:413-418

In: Chemical Reactions in Liguid and Solid Phase
Editors: G. E. Zaikov and A. Jiménez, pp. 119-133
ISBN 1-59033-854-5

Chapter 5

MONOMOLECULAR CHAIN TERMINATION IN KINETICS OF DIMETHACRYLATES POSTPOLYMERIZATION

***Yu. G. Medvedevskikh*[1]*, A. R. Kytsya*[1]*, L. I. Bazylyak*[1]*, A. M. Bratus*[1]*, A. A. Turovski*[1] *and G. E. Zaikov*[2]**

[1]Physical Chemistry and Technology of Combustible Minerals Department. L. V. Pisarzhevsky Institute of Physical Chemistry, National Academy of Sciences of Ukraine. 3a Naukova Str., 9053, Lviv, UKRAINE
e–mail: vfh@org.lviv.net

[2]N.M. Emanuel Institute of Biochemical Physics. Russian Academy of Sciences. 4 Kosygin Str., 119991, Moscow, RUSSIA
e–mail: chembio@sky.chph.ras.ru

ABSTRACT

Kinetics of postpolymerization (after *UV*–illumination stopping) for a number of dimethacrylates differed by nature and molecular mass has been experimentally studied for a wide range of temperatures. A series of kinetic curves differed by starting conversion of the dark period of time has been obtained for every value of the temperature. Proposed kinetic model of process is based on the following main principles, namely: the process at an interface on the boundary "liquid monomer – solid polymer (micrograins)" takes a main share into kinetics of postpolymerization; chain termination at an interface is monomolecular and is controlled by chain propagation rate and represents by itself the self–burial act of active radical in the conformation trap; monomolecular chain termination is characterized by a wide spectrum of characteristic times and that's why function of the relaxation is described by Kohlrausch's stretched exponential law.

Obtained kinetic equation is in good agreement with the all set of experimental data. This permitted to estimate the constants (k_t) rate of chain termination and to determine the scaling dependence of k_t upon molar–volumetric concentration $[M_0]$ of monomer in bulk.

It was assumed, that the stretched exponential law and scaling dependence k_t from $[M_0]$ are characterized by common peculiarities, namely by a wide range of characteristic times of relaxation possessing by property of fractal set.

Key words: monomolecular chain termination, dimethacrylates, postpolymerization

1. Introduction

Postpolymerization or, so–called, non–stationary dark process of polymerization proceeding after *UV*–illumination stopping gives important information about chemical mechanism of chain termination elementary reactions. Experimental data detect a series of general regularities [1–4] both for mono– and multifunctional monomers, namely: 1) rate and polymerization degree appreciably depend on starting conversion of dark period; 2) number–average molecular mass of polymer is propagated via polymerization process proceeding and at the same time concentration of radicals detected by *EPR*–spectroscopy "*in situ*" is insignificantly decreased or is constant; 3) there are two sections on the kinetic curve, namely first is rapid and short, the second is slow and long. All above–said permit to suppose the fact that radicals initiating the postpolymerization process are sharply distinguished by characteristic times of life. Maybe that is why three variants of postpolymerization kinetics description based upon bimolecular, linear and mixed chain termination checked in the work [5] show, that kinetic models *only* with bimolecular or linear chain termination are less satisfactory agreed with experimental data than their combination representing by itself mixed chain termination and permitting to introduce at least two characteristic times of radical's life. It is supposed also in the work [5], that postpolymerization process proceeds with the same rate in the all volume of polymerizing system.

The other way round, kinetic models of stationary and non–stationary (dark) processes of photoinitiated polymerization proposed in works [6–8] are based on the imagination about microheterogeneity of polymerizing system and its two reactive zones. That is why observed rate of polymerization ω is a sum both of the rate of homophaseous process proceeding in the volume of liquid monomer in accordance with the classic kinetic scheme with the bimolecular chain termination and heterophaseous process proceeding in the interface layer on the boundary "liquid monomer – solid polymer (micrograins)" under gel–effect condition with the monomolecular chain termination $\omega = \omega_v\varphi_v + \omega_m\varphi_m$, where φ_v and φ_m are volumetric parts of liquid monomer phase and interface layer and ω_v and ω_m are corresponding specific rates of polymerization.

However, it is taken into account at the derivation of kinetic model for the postpolymerization that radicals in liquid phase are characterized with little time of "life" and can not take remarkable share into postpolymerization process. So, that is why the rate of postpolymerization is completely determined by the rate of heterophaseous process in the interface layer: $\omega \cong \omega_m\varphi_m$. With the aim of taken into account the difference between characteristic times of radicals "life", kinetic scheme of process in the interface layer has been proposed in [8] as primary (1) and secondary (3) chains with the monomolecular chain termination and two characteristic times, namely $\tau_1 = (k_t [M_m])^{-1}$ and $\tau_2 = (k_t'[M_m])^{-1}$:

$$R_m + M_m \begin{cases} \xrightarrow{k_p} R_m \\ \xrightarrow{k_t} R_z \end{cases} \qquad (1)$$

$$R_z + M_m \xrightarrow{k_m} R'_m \qquad (2)$$

$$R'_m + M_m \begin{cases} \xrightarrow{k_p'} R''_m \\ \xrightarrow{k_t'} R'_z \end{cases} \qquad (3)$$

The main peculiarity of schemes (1) and (3) is fact, that the rate of monomolecular chain termination is controlled by rate of chain propagation since acts of chain propagation and chain termination represent by themselves two different results of interactions of active radicals R_m and R_m' with functional group of monomer leading correspond to formation again of active radicals (so–called chain propagation) or their "freezing" [3, 9], trapping [4, 10] (monomolecular chain termination), *i.e.* leading to formation of non–active radicals R_z and R_z'. Elementary reaction (2) represents by itself the reaction of secondary active radicals R_m' initiation from the primary frozen R_z one's.

Although kinetic scheme (1) – (3) significantly idealized, it can be explained from the physical point of view. This physical sense is based on the generality of kinetic regularities of polymerization at an interface (including specifically the first order upon initiator) and, for example, bulk polymerization [11], polymerization in the presence of porosity fillers [12] or at the adsorbed immobilization on the surface of carrier of polymeric initiator [13]. This generality means that solid phase creates the especially ordered structure of the nearest reactive space in which translational and segmental mobility of macroradicals are sharply decreased. Such structure is similar to system of weakly interacted between each other microreactors [11–15] in every of which chain termination is determined by the rate of chain propagation in such manner that the decay of radicals represents by itself their self–burial act [8].

Proposed on the base of kinetic scheme (1) – (3) equation of the postpolymerization absolutely well describes experimental kinetic curves. This permitted for the first to estimate numerical values of constants rate for monomolecular chain termination ($k_t \sim 10^{-5}$ $m^3/mole\ s$, $k_t' \sim 10^{-6}$ m^3/mole s) and to ascertain scaling dependencies $k_t \sim [M_0]^{-m}$, $k_t' \sim [M_0]^{-n}$, $m = 1.5$, $n = 2.5$, where $[M_0]$ is concentration of monomer in bulk [16]. Above–mentioned dependencies do not agree with the kinetic scheme (1) – (3) with two characteristic times of macroradicals life, since indicate upon wide spectrum of characteristic times. Besides, with the aim of the assumption of fact, that the self–burial act of active radical represents by itself the act of chain propagation, is needed at least to estimate the activated nature k_t. That is why in the presented work the temperature dependence of postpolymerization rate for a range of dimethacrylates has been experimentally studied and a stretched exponential law (in accordance with [17]) has been used for description of kinetics of process.

2. Experimental

Kinetics of postpolymerization of dimethacrylates, namely *MGPh–9* $\{M\text{–}O\text{–}(CH_2CH_2O)_3\text{–}C(O)\text{–}C_6H_4\text{–}C(O)\text{–}(OCH_2CH_2)_3\text{–}O\text{–}M\}$, *OCM–2:* $\{M\text{–}O\text{–}(CH_2CH_2O)\text{–}C(O)\text{–}(CH_2\text{–}CH_2O)_2\text{–}C(O)\text{–}O\text{–}(CH_2CH_2O)\text{–}O\text{–}M\}$, *TGM–3:* $\{M\text{–}O\text{–}(CH_2\text{–}CH_2\text{–}O\text{–})_3\text{–}M\text{–}\}$ and *DMEG*: $\{M\text{–}O\text{–}CH_2\text{–}CH_2\text{–}O\text{–}M\text{–}\}$, where M is $CH_2{=}C(CH_3)\text{–}C(O)\text{–}$ has been studied in a range of 5 - 50 ^0C, and also 1,4–butandioldimethacrylate *(BDDMC)* at 20 ^0C.

Experiments have been carried out on the interferometric laser plant measuring current (H) and limited achieved (H_0) contractions of the layer of photopolymerizing composition. When taking into account the ratio of above–mentioned contractions, the relative degree of polymerization (or conversion) was determined, namely $P = H/H_0$. Integral *UV*–illumination of the *DRT*–400 lamp at the intensity of falling upon surface of photopolymerizing composition luminous flux consisting of 37.4 W/m^2 has been used for carrying out of experiments. *UV*–illumination was stopped at the defined stage of luminous polymerization process prolonging to register the contraction of composition layer under dark conditions; after that *UV*–illumination was switched on providing in this manner the finish of the polymerization process proceeding with the determination of the final contraction H_0. With taken into account the above–mentioned value H_0, the thickness of the layer has been determined. Above–said thickness of the photocomposition layer was varied in a ranges of $(0.5 \div 3.0) \times 10^{-4}$ m. 2,2–Dimethoxy–1,2–diphenylethane–1–one has been used as photoinitiator of the process at its starting concentration 1 % by mass.

Typical experimental kinetic curves of the postpolymerization are represented on *Figure* 1 as dependence of conversion $P\text{–}P_0$ increment from time t of dark period.

3. Results and Discussion

Let us limit by only kinetic scheme (1) for description of postpolymerization kinetics. In accordance with the above–mentioned scheme rate of chain propagation (w_p) and rate of monomolecular chain termination (w_t) are determined by following equations:

$$w_p = k_p\,[M_m]\,[R_m] \tag{4}$$

$$w_t = k_t\,[M_m]\,[R_m] \tag{5}$$

So, the rate of radical's concentration changing under dark conditions is equal to

$$d[R_m]/dt = -\,k_t\,[M_m]\,[R_m] \tag{6}$$

Assuming that $[M_m] \approx const$, we obtain

$$[R_m] = [R_{m0}]\,exp\,\{-\,\beta\,t\}, \tag{7}$$

where $[R_{m0}]$ is starting concentration of radicals in the reactive zone for dark period of time and parameter

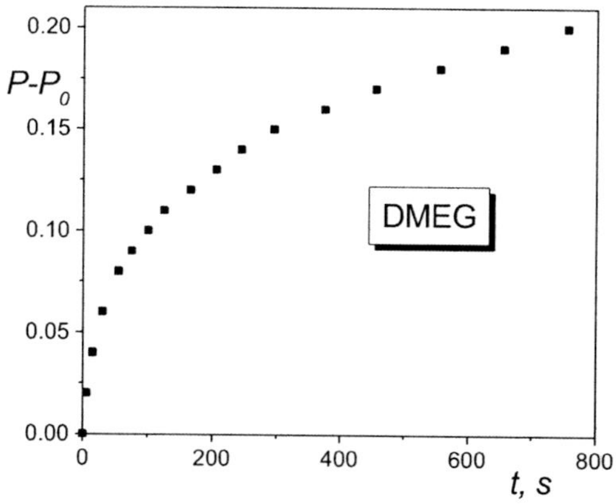
0.20
P-P0
0.15
0.10
0.05
0.00
0
200
400
600
800
t, s
DMEG

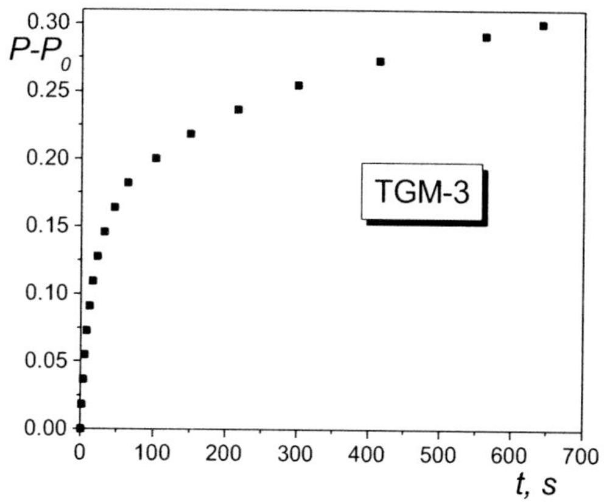
0.30
P-P0
0.25
0.20
0.15
0.10
0.05
0.00
0
100
200
300
400
500
600
700
t, s
TGM-3

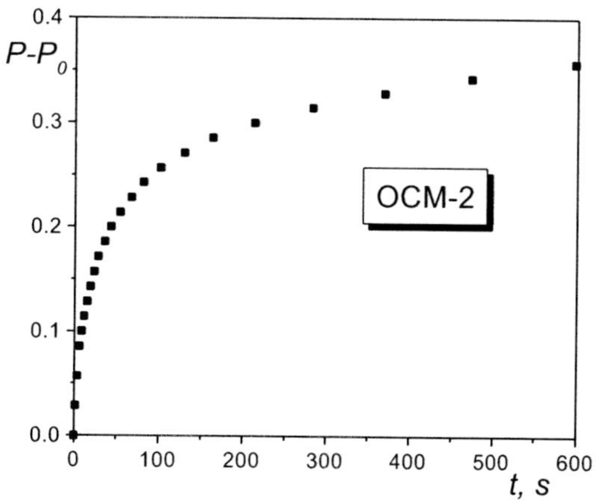
0.4
P-P0
0.3
0.2
0.1
0.0
0
100
200
300
400
500
600
t, s
OCM-2

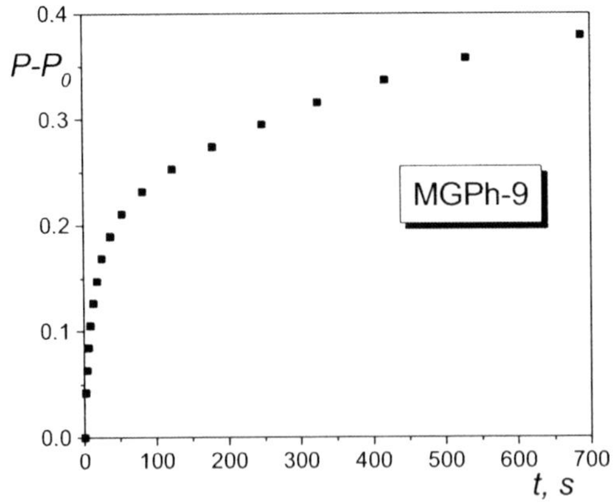

Figure 1. Typical experimental kinetic curves of dimethacrylates postpolymerization at 15 ^0C"

$$\beta = k_t\,[M_m] \approx \tau_t^{-1} \tag{8}$$

determines characteristic time τ_t of radical's life or relaxation time.

In accordance with the starting principle [**8**] we assume, that the process at an interface takes the main share in kinetics of postpolymerization resulting the

$$-d[M]/dt = w_m\varphi_m = w_p\varphi_m \tag{9}$$

where $[M]$ is molar–volumetric concentration of monomer calculated upon whole volume of system.

Volumetric part φ_m of the interface layer is approximated by function [6–8]

$$\varphi_m \approx h\,(F_s/F_v)\,\varphi_s\,(1-\varphi_s) \approx h\,(F_s/F_v)\,P(1-P) \tag{10}$$

where φ_s is volumetric part of solid polymeric phase, h is thickness of the interface layer, F_s/F_v is ratio of fractal characteristic of surface and volume of micrograins of the solid polymeric phase into liquid monomeric one or microdrops of liquid monomeric phase into solid polymeric matrix.

By combining (4), (7) and (10) into (9) and expressing the rate of process in the units of conversion $dP/dt = -(d[M]/dt)/[M_0]\Gamma_0$, where $[M_0]$ is concentration of monomer in bulk and Γ_0 is the final degree of double bond conversion in the solid polymeric phase, we obtain

$$dP/dt = k_p[M_m][R_{mo}]\,h\,(F_s/F_v)\,P(1-P)exp\{-\beta t\} \tag{11}$$

At $t = 0$ we have $P = P_0$ and

$$(dP/dt)_{t=0} \equiv w_0 = k_p[M_m][R_{mo}]\, h\, (F_s/F_v)\, P_0(1 - P_0) \tag{12}$$

represents by itself the starting rate of postpolymerization for dark period of time.

With taken into account (12), equation (11) for determination of postpolymerization rate can be represented as

$$\frac{dP}{dt} = w_0 \frac{P(1-P)}{P_0(1-P_0)} exp\{-\beta t\} \tag{13}$$

As we have expected equation (13) with single characteristic time $\tau_t = \beta^{-1}$ satisfactory describes only the starting rapid and short section of experimental kinetic curve. This means that monomolecular chain termination is characterized with the wide spectrum of relaxation time. That is why the stretched exponential law has been used for full description of the kinetic curve. In accordance with this law the function of relaxation is writing as follows:

$$\psi(t) = A\, exp\{-bt\}^{\gamma},\ 0 < \gamma < 1 \tag{14}$$

Namely value $0 < \gamma < 1$ gives the effect of the stretched exponential to (14).

Stretched exponential law in the form (14) has been proposed by Kohlrausch (see [17]) firstly in 1864 for the description of materials creeping and has been used by Williams and Watts [18] in 1970 at the analysis of dielectric relaxation of polymers. It was disclosed from this time, that above–mentioned law describes the different types of relaxation mainly in the non–ordered systems in which spatial inhomogeneity forms the hierarchy of relaxation times subordinating to properties of fractal set [19].

By setting the effect of stretched exponential into (13) we will obtained:

$$\frac{dP}{dt} = w_0 \frac{P(1-P)}{P_0(1-P_0)} exp\{-\beta t\}^{\gamma} \tag{15}$$

Integral form of the equation (15) has not the simple analytic view, and that is why integral kinetic curves $P–P_0 = f(t)$ (see Figure 1) were numerically differenced for comparison of (15) with experimental data. After this, differential kinetic curves $dP/dt = f(t)$ have been obtained; typical examples of such curves represented on Figure 2 by points.

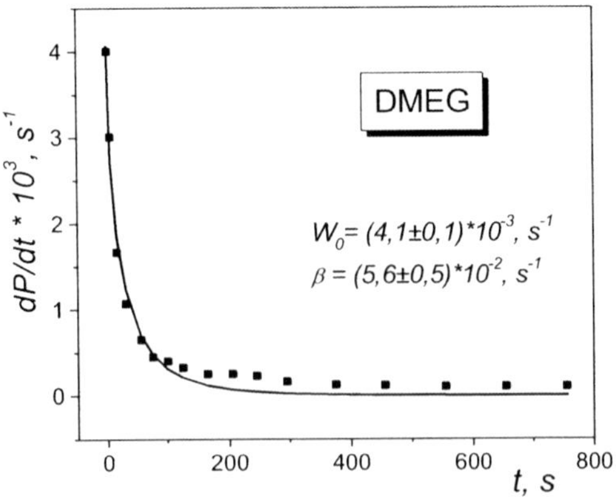
DMEG
W_0= (4,1±0,1)*10^{-3}, s^{-1}
β = (5,6±0,5)*10^{-2}, s^{-1}
dP/dt * 10^3, s^{-1}
t, s
0
200
400
600
800

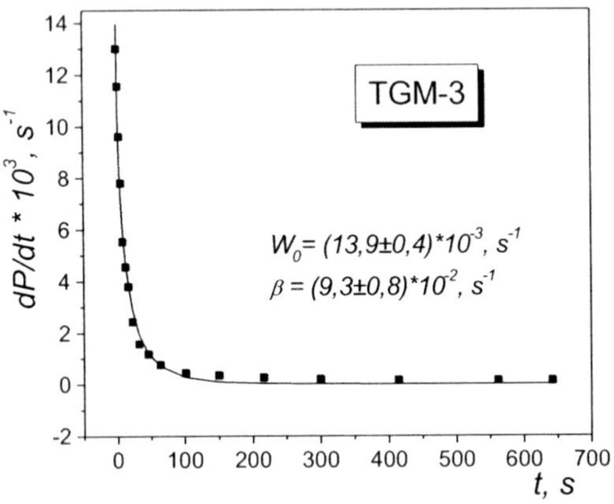
TGM-3
W_0= (13,9±0,4)*10^{-3}, s^{-1}
β = (9,3±0,8)*10^{-2}, s^{-1}
dP/dt * 10^3, s^{-1}
t, s
0
100
200
300
400
500
600
700

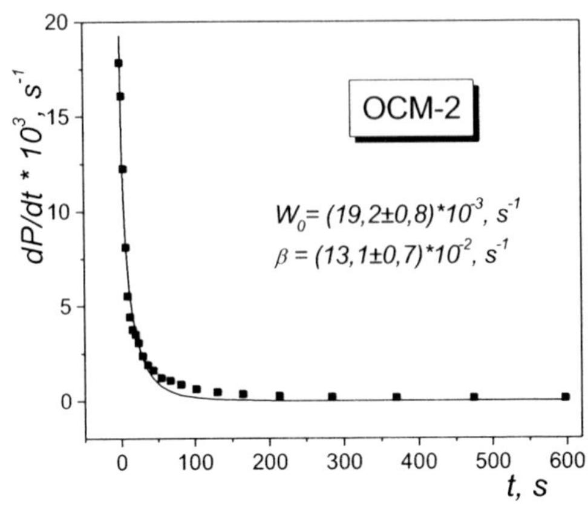
OCM-2
W_0= (19,2±0,8)*10^{-3}, s^{-1}
β = (13,1±0,7)*10^{-2}, s^{-1}
dP/dt * 10^3, s^{-1}
t, s
0
100
200
300
400
500
600

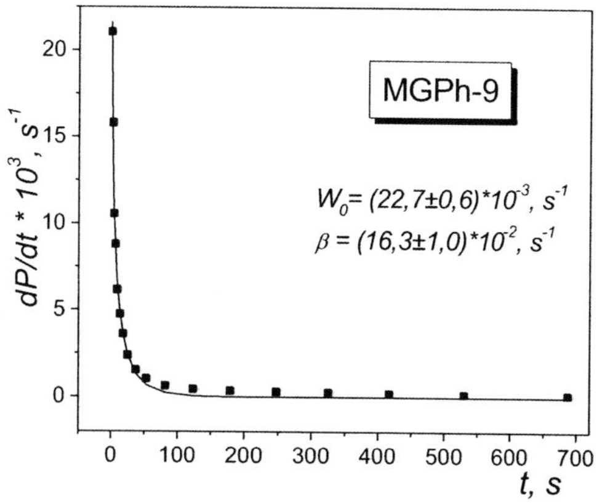

Figure 2. Interpretation of experimental kinetic curves of dimethacrylates postpolymerization in the differential form (points) and compared with calculated one's via equation (15) (lines) at 15 ^0C.

On the basis of comparison of above–said obtained curves and equation (15) by optimization method the all three constants of the equation (15) have been found, namely w_0, β and γ. As was discovered, the spread in numerical values for parameter γ from 0.5 to 0.7 was the less satisfactory at the optimization selection of all parameters in the experimental series. This can be explained in fact, that the most important section of the kinetic curve for estimation of parameter γ, namely slow and long is characterized by low values dP/dt and takes a little share into the functional of number squared of calculated deviations from experimental ones. Fixing the parameter γ showed, that the minimal standard error at the estimation of w_0 and β is observed at the value $\gamma = 0.6$. That's why all experimental curves $dP/dt = f(t)$ were compared with the equation (15) at fixed value $\gamma = 0.6$ with optimization only upon two parameters, namely w_0 и β. Examples comparison of calculated (lines) in accordance with the equation (15) and experimental (points) kinetic curves is represented on *Figure* 2. It can be notified, that in the all cases satisfactory correspondence takes place between experimental and calculated accordingly to equation (15) kinetic curves.

As was mentioned earlier [6, 7], let us notify that experimental error at the construction of separate kinetic curve of postpolymerization is sufficiently small. This can be proved by points location on the Figure 1. At the same time, spread in kinetic curves under the same conditions of experiment proceeding considerably exceeds the error of the individual kinetic curve. This phenomenon is well–known as a bad reproducing the kinetic measurements and, in accordance with the author's mind [6, 7] represents by itself the direct proof of the microheterogeneity of polymerizing system, fluctuated mechanism of the formation and propagation of solid polymeric phase and interface layer on the boundary of "liquid monomer–solid polymer" and also its active role via postpolymerization. That's why from 3 till 10 kinetic curves have been obtained for each condition of experiment. Founded from these curves kinetic parameters of process have been averaged. In such a way the statistical truth of obtained estimations was stepped up. For example, we can see in Table 1 values of the parameter β obtained from individual kinetic curves and also results of their average. We can see from Figure 3 that in spite of the spread in values of the parameter β, these values do not depend from the starting conversion P_0 of dark period. The same picture was observed for another conditions of the experiment proceeding.

Table 1. Nature of Spread for Parameters W_0 and β Calculated via Equation (15) for Postpolymerization of DMEG at t = 15 ^{0}C

№	P_0	τ, s	$W_0 \times 10^3 . s^{-1}$	$\beta \times 10^2 . s^{-1}$
1	0.30	650	4.0±0.1	5.7±0.4
2	0.29	700	3.7±0.1	4.6±0.3
3	0.16	650	3.4±0.1	7.4±0.7
4	0.22	750	4.1±0.1	5.6±0.5
5	0.19	700	3.6±0.2	5.5±0.6
6	0.21	750	3.2±0.1	5.3±0.6
$\overline{\beta} \times 10^2 . s^{-1}$				5.7±0.9

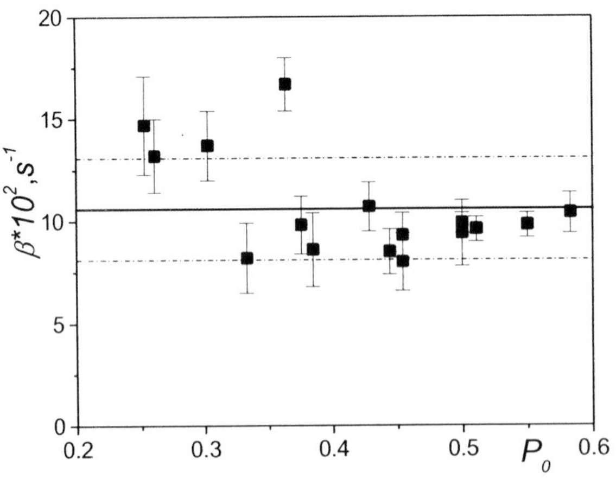

Figure 3. Calculated via optimization method values of the β parameter at different starting conversions of postpolymerization for dimethacrylate *TGM*–3 at 15 ^{0}*C*.

Average values β for investigated dimethacrylates at different temperatures are represented in Table 2. Interpretation of these data in Arrhenius coordinates is shown on Figure 4; corresponding linear equations of regression are presented on the same Figure 4. Activation energies of monomolecular chain termination constants rate have been calculated in accordance with above–mentioned equations and presented in Table 3.

Table 2. Average Values of Parameter β Accordingly to Kinetic Equation (15) Calculated upon Experimental Data for Dimethacrylates Postpolymerization at Different Temperatures

t. ^{0}C	$\beta \times 10^2$. s^{-1}				
	DMEG	BDMC	TGM–3	OCM–2	MGPh–9
5	4.6±2.3	–	10.7±3.0	10.9±3.0	–
10	–	–	10.7±2.1	13.3±2.1	12.8±1.2
15	5.7±0.9	–	10.6±2.5	15.3±2.1	15.6±2.8
20	–	9.8±7.1	12.2±0.7	15.0±2.0	13.8±3.4
25	6.4±2.8	–	14.6±4.5	13.8±2.6	16.5±2.3
30	–	–	–	–	17.6±2.2
35	–	–	14.3±4.1	16.5±4.0	19.7±7.4
40	–	–	–	–	19.0±4.4
45			–	–	19.7±5.2
50	–	–	14.9±4.0	19.1±2.8	17.6±6.1

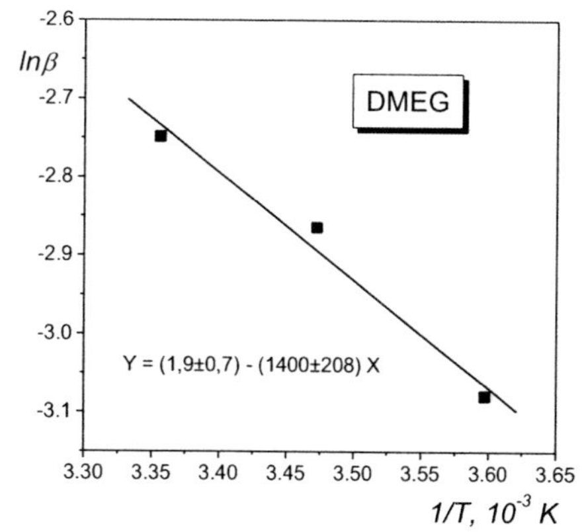

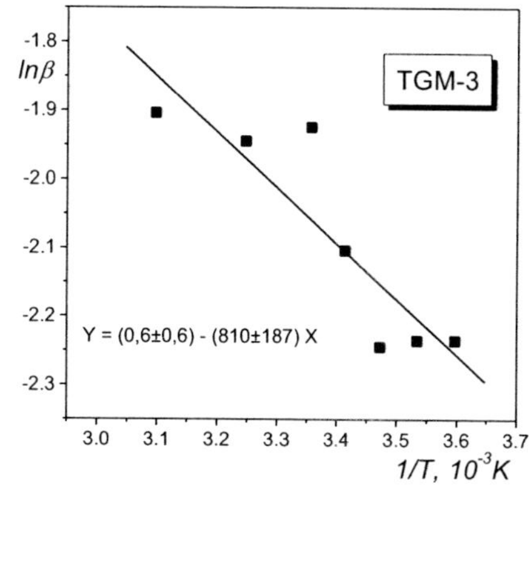

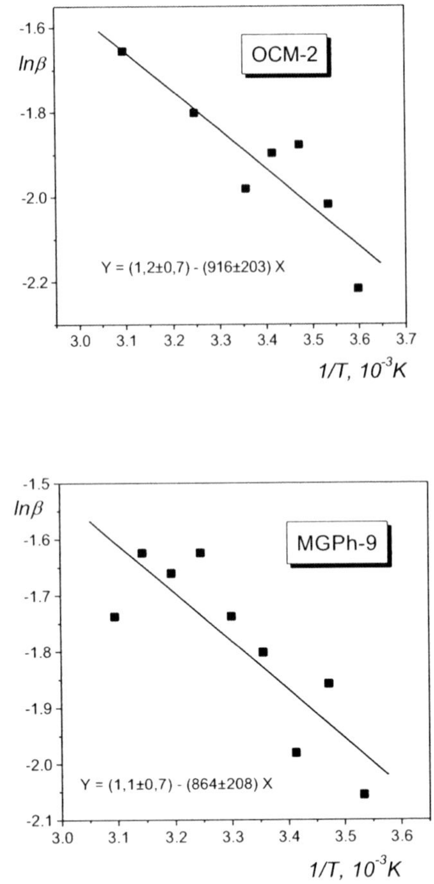

Figure 4. Temperature dependencies of the β parameter in the coordinates of the Arrhenius equation.

Table 3. Kinetic Constants of Dimethacrylates Postpolymerization

Dimethacrylate	*M*, *kg/mole*	*[M]*$_0$, *mole/m*^{3}	$\overline{\beta}\times10^2$, s^{-1}	$\overline{k_t}\times10^5$, $m^3/mole\times s$	E_A, *kJ/mole*	$k_p\rho_0^{\xi}$
DMEG	0.198	5297	6.2	1.17	11.6	19.4
BDMC	0.226	4526	9.8	2.16	–	27.6
TGM–3	0.286	3814	12.2	3.20	6.7	30.6
OCM–2	0.418	2890	14.6	5.05	7.6	30.4
MGPh–9	0.566	2060	15.1	7.33	7.2	25.1

Note: Values $\overline{\beta}$ were calculated at t = 20 ^0C via linear equations of regression of temperature dependence (see *Figure* 4): $\overline{k}_t = \overline{\beta}/[M_0]$; $k_p\rho_0^{\xi} = \overline{k}_t\rho_m^{\xi} = \overline{k}_t[M_0]^{\xi}$, where ξ = 1.67.

As we can see from presented data, activation energies of monomolecular chain termination constants rate for three dimethacrylates (namely, *OCM*–2, *MGPh*–9 and *TGM*–3) are similar between each other and equal to 6.7 – 7.6 *kJ/mole*; at the same time this value for *DMEG* is 11.6 *kJ/mole*. So, monomolecular chain termination represents by itself the activated process and low values of activation energies are considerably less than activation energies of viscous flow (for example, for *OCM*–2 and *MGPh*–9 these values are equal to ≈ 55 and 44 *kJ/mole* correspondently) and do not agree with the accepted assumption accordingly to which the monomolecular chain termination rate is controlled by its propagation rate, and, as follows from this, monomolecular chain termination can not be considered as passive "freezing" of active radical.

With the aim of decreasing the element of randomness at the estimation of parameter β at the interpretation of scaling dependence β and k_t from $[M_0]$, values β at temperature 20 ^0C also have been calculated on the basis of linear equations of regression. On the basis of these values $\beta k_t = \beta/[M_0]$ at 20 ^0C have been calculated with the assumption that concentration of monomer in the interface layer is near to its concentration in the bulk, namely $[M_m] \cong [M_0]$. Obtained results are also represented in Table 3.

At the comparison of obtained in the presented work from kinetic model with stretched exponential law values of k_t and obtained earlier one's in the work [16] accordingly to kinetic scheme with two characteristic times and exponential law of relaxation, it can be notified that they are practically coincided. For example, values of k_t for *BDMC* are 2.2 and 1.4×10^{-5} $m^3/mole\times s$, for *OCM*–2 are 5.0 and 3.6×10^{-5} $m^3/mole\times s$, for *MGPh*–9 are 7.3 and 5.2×10^{-5} $m^3/mole\times s$ correspondently. So, both kinetic models in spite of their difference lead to corresponding estimations of k_t. This means, that the main peculiarity of dimethacrylates postpolymerization is monomolecular chain termination accordingly to scheme (1).

Interpretation of scaling dependence of calculated values β from concentration of monomer in bulk $[M_0]$ is represented on Figure 5. If we designate this dependence as $\beta \sim [M_0]^{1-\xi}$, than in accordance with linear equation of regression (see Figure 5) we will obtain $1-\xi = -0.86 \pm 0.27$. As follows from this, scaling dependence $k_t \sim [M_0]^{-\xi}$ and $\xi = 1.86 \pm 0.27$ takes place for constant rate of monomolecular chain termination.

As it was above–said, scaling dependence $k_t \sim [M_0]^{-\xi}$ has been determined earlier in the work [16] despite the fact, that postpolymerization kinetics in this work described on the basis of kinetic scheme (1) – (3) with two characteristic times of relaxation. This fact means, that scaling dependence $k_t \sim [M_0]^{-\xi}$ is determining by fundamental factors but not by choice of postpolymerization kinetics equation (exponential law with two characteristic times of relaxation based upon scheme (1) – (3) or stretched exponential law based only upon scheme (1)). But, at the same time stretched exponential law demands the spectrum of characteristic times of relaxation possessed by properties of fractal set [19]. Scaling form $k_t \sim [M_0]^{-\xi}$ is also directed upon this fact. So, we conclude that the relation between scaling form and stretched exponential law should be exist. Maybe, that's why $1/\xi \approx \gamma$, but, however, it's needed the special theoretical discussion.

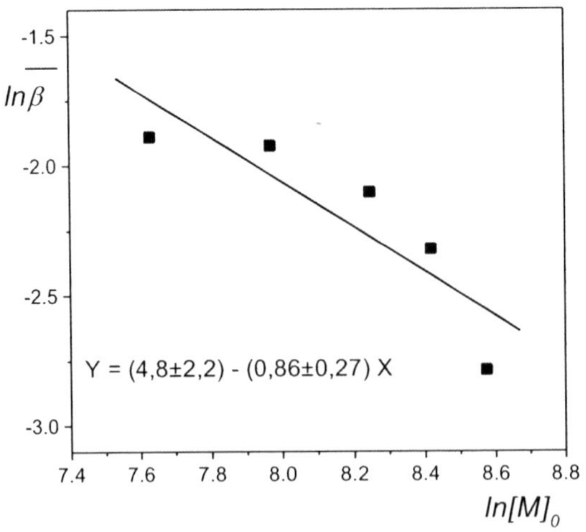

Figure 5. Dependence of calculated values of the $\overline{\beta}$ parameter at $t = 20\ ^0C$ upon molar–volumetric concentration $[M_0]$ of dimethacrylate in bulk.

ACKNOWLEDGEMENT

This work has been performed under financial support of the STCU (project №1447)

REFERENCES

[1] Garina E. S., Kuznietsova T. M., Zubov V. P., Kabanov V. A. // *Dokl. AN USSR* (1973), v. 209, № 2, P. 380.

[2] Shen J., Tian Y., Wong G. // *Sci. China* (1990), v. 33, № 9, P. 1046.

[3] Dýachkov A. N., Efimov A. L., Efimov A. L. *ect.* // *Vysokomol. soed., ser. A* (1983), v. 25, № 10, P. 2176.

[4] Zhu S., Tian Y., Mamieles A. E. // *Macromolecules* (1990), v. 23, P. 1144.

[5] Andrzejevska E., Bogacki M. B., Andrzejevski M. // *Macromol. Theory Simul.* (2001), v. 10, P. 842.

[6] Medvedevskikh Ju. G., Zagladko E. A., Turovski A. A., Zaikov G. E. // *Int. J. Pol. Matter.* (1999), v. 43, P. 157.

[7] Zagladko E. A., Medvedevskikh Ju. G., Turovski A. A., Zaikov G. E. // *Int. J. Pol. Matter.* (1998), v. 39, P. 227.

[8] Medvedevskikh Ju. G., Zagladko E. A., Turovski A. A., Zaikov G. E. // *Rus. Polym. News* (1999), v. 4, № 3, P. 33.

[9] Efimov A. L., Bugrova T. A., D'yachkov A. I. // *Vysokomolek. soed., ser. B* (1992), v. 25, № 11, P. 2176.

[10] Kurdikar D. L., Peppas N. A. // *Macromolecules* (1981), v. 27, P. 4084.
[11] Ivanov V. B., Romaniuk A. P., Shybanov V. V. // *Vysokomolek. soed., ser. B* (1993), v. 35, № 2, P. 119.
[12] Min'ko S. S., Luzinov I. A., Smirnov B. Ph. // *Vysokomolek. soed., ser. B* (1990), v. 32, № 10, P. 750.
[13] Min'ko S. S., Sydorenko A. A., Voronov S. A. // *Vysokomolek. soed., ser. B* (1995), v. 37, № 8, P. 1403.
[14] Ivanchev S. S., Dmitrenko A. B., Krupnyk A. M. // *Vysokomolek. soed., ser. A* (1998), v. 30, № 9, P. 1951.
[15] Stepanian A. O., Zaremsky M. Yu., Olenin A. V. // *Dokl. AN USSR* (1984), v. 274, № 3, P. 655.
[16] Medvedevskikh Yu. G., Zaglad'ko E. A., Bratus' A. N., Turovskyj A. A. // *Dokl. AN USSR* (2000), № 10, P. 148.
[17] Montroll E. W., Bendler J. T. // *J. Stat. Phys.* (1984), v. 34, P. 129.
[18] Williams G., Watts D. C. // *Trans. Faraday Soc.* (1970), v. 66, P. 80.
[19] Shlezinger M., Klafter J. // *Abstracts of the International Symposium on Fractals in Physics*, Moscow, Myr (1988), P. 553.

In: Chemical Reactions in Liguid and Solid Phase
Editors: G. E. Zaikov and A. Jiménez, pp. 135-143
ISBN 1-59033-854-5

Chapter 6

THERMODYNAMICS OF MIXING OF CRYSTALLINE POLY(ETHYLENE OXIDE) WITH AMORPHOUS POLYMERS OF VINYL SERIES

Anna I. Suvorova, Irina S. Tyukova and Albina H. Hassanova
Ural State University,
Ekaterinburg, Lenin Ave.51,
620083, Russia

ABSTRACT

The interaction of crystalline poly(ethylene oxide) (PEO) and its blends with amorphous poly(2-vinyl pyridine) (P2VP), poly(4-vinyl pyridine) (P4VP), poly(methyl acrylate) (PMA), poly(methyl methacrylate) (PMMA), poly(butyl acrylate) (PBA) and poly(butyl methacrylate) (PBMA) with chloroform was investigated using the static interval sorption method. Calculations of the Gibbs energy of polymer mixing with solvent has been calculated taking into account the degree of crystallinity of the blends as determined from X-ray diffraction data. The energy of mixing of crystalline PEO with amorphous polymers was calculated from the obtained Gibbs energies of mixing with chloroform using the thermodynamic cycle. These energies is changed in the range from –2.3 to –9.7 kJ/kg depending on the structure of amorphous components.

Keywords: Thermodynamics; Gibbs energy; blends; poly(ethylene oxide); amorphous polymers.

INTRODUCTION

The thermodynamics of interaction between the components of polymer blends remains a point of research interest, regardless of the fact that compatible and incompatible systems and phase states of blends of polymers of different classes have already been systematized in sufficient detail [1–6]. The specific features of the phase state of the blends of crystalline and amorphous polymers and, particularly, the thermodynamics of their mixing are less known; the properties of such systems are considered, as a rule, in a narrow range of compositions and for a single amorphous polymer introduced into the polymer crystalline matrix. In this work the thermodynamics of mixing of crystalline poly(ethylene oxide) with amorphous polymers of the vinyl series such as: poly(vinyl pyridines), poly(acrylates), poly(methacrylates) were investigated. Poly(ethylene oxide) (PEO), which possesses a high crystallinity, is of practical interest owing to its use in the design of different thin film materials for sensors and electrochemical devices [7-10]. Previously it was shown that the introduction of amorphous polymers of series poly(acrylates) and poly(methacrylates) into crystalline PEO leads to a change in its melting temperature and structure [11-12].

To elucidate the role of the compatibility of components and features of their chemical molecular structure in the formation of the submicroscopic structure of the blends, we estimated the Gibbs energy of mixing (g_x) of the crystalline polymer PEO with some amorphous polymers of the vinyl series, because the magnitude of Δg_x and its sign characterize the possibility of spontaneous mixing of the components.

EXPERIMENTAL

Commercial PEO and acrylic polymers, such as poly(methacrylate) (PMA), poly(butyl acrylate) (PBA), poly(methyl methacrylate) (PMMA), poly(butyl methacrylate) (PBMA) and poly(2-vinylpyridine) (P2VP), poly(4-vinylpyridine) (P4VP) were used as the research subjects. All commercial polymers were purified by precipitation from chloroform solutions into ethanol, dried in air and then in vacuum at 25 °C. The characteristics of the examined polymers are listed in Table 1.

Table 1. Characteristics of the Tested Polymers

Polymer	Molecular weight, M_η *	Molar volume of chain unit. $V_{mol}\,10^3$, m^3/kmol	Glass transitions temperature, T_g, °C	Melting temperature, T_m, °C
PEO	1×10^6	39.2	-67	64
PMA	3×10^4	80.4	8	
PBA	2×10^5	117.8	-54	
PMMA	9×10^5	84.0	104	
PBMA	1×10^6	134.6	20	
P2VP	2×10^5	100.9	170	
P4VP	2×10^5	100.9	172	

* M_η was determined by intrinsic viscosity of the chloroform solutions of all polymers.

In order to follow the influence of the molecular structure of the repeating units of polymer chain of amorphous components on the properties of PEO, we prepared polymer blends of PEO and poly(acrylates), poly(methacrylates) with the same molar fraction (per mole of repeating unit) of crystalline component to be equal to 0.93 (80 -87 mass%, depending on the molecular mass of the repeating unit of an amorphous polymer). Poly(vinyl pyridines) had equal molar volume of their chain unit and they were induced into the blends in two content of 10 and 25 mass%. PEO – amorphous polymer films were obtained by pouring out homogeneous 1–3% chloroform solutions of polymer blends onto a Teflon dish. After chloroform was evaporated at 25°C in air, then films were dried to a constant weight in a vacuum at residual pressure of 10 Pa at the same temperature. The films were 50–60 μm thick and stored before study in desiccators. The degree of crystallinity of PEO and all polymer blends was estimated from the wide angle X-ray diffraction data. Diffraction patterns of polymer films were recorded on a DRON-2.0 goniometer (CuK_α radiation, λ=0,1542 nm) in the range 5°– 42° of angles (2 θ). The degree of crystallinity (α) was calculated as the ratio of the sum of areas under "crystalline" peaks in the X-ray scattering curve to the total area under the curve in the range 2 θ = 15°–35° minus the background. The degree of crystallinity of PEO and PEO blends is listed in the third column of the Table 2.

The amount of the chloroform vapor absorbed by polymer films at 25 °C was determined by the weight variant of isothermal interval sorption method [13]. We used high vacuum apparatus providing 10^{-2} Pa of residual pressure with a Mc Bain quartz helix microbalance with sensitivity of 0.3–0.5 m/kg. The helix extension was measured accurate to 5×10^{-6} m using a V-630 cathetometer. The error in determination of chloroform sorption was no greater than 2%, the chloroform vapor pressure was measured by an OM-2 pressure gauge with an error no greater than 0.1%.

RESULTS AND DISCUSSION

The sorption isotherms of chloroform vapor by individual PEO, poly(vinyl pyridines) and their blends in absorbed amount (x/m) – equilibrium relative vapor pressure (p_1/p_1^0) coordinates are presented in Figure 1. Similar isotherms of chloroform vapor for the PEO - poly(acrylates) and poly(methacrylates) systems were received early [13].

The sorption isotherms have typical shape for polymer solution: they are concave toward the axis of coordinates, which is characteristic of closely packed polymers with unlimited swelling in the solvent [14]. Throughout the entire range of p_1/p_1^0 the sorption capacity of poly(vinyl pyridines), which are in the glassy state at 25 °C, is small. The sorption isotherms of PEO substantially differ from those for the second component of the blends: in the initial portion of the isotherm, the sorption capacity of PEO is very small, which is characteristic of crystalline polymers [15, 16]. In the range of relative pressures $p_1/p_1^0 \sim$ 0.5-0.7, the sorption isotherm for the PEO – chloroform system has a knee, behind which the solvent sorption sharply increases. The same behavior of the sorption isotherm of chloroform vapor by crystalline PEO was found in [16]; this behavior is associated with the isothermal melting of polymer crystallites during sorption. The isotherms behave of the blends are analogously to the isotherm for PEO and only slightly deviate from the isotherm of the crystalline polymer throughout the entire range of relative pressures of chloroform vapors. More detailed consideration of the sorption data shows that, at low chloroform vapor pressures, the sorption capacity of blends of PEO is lower than for individual PEO. It is well known that the sorption of solvent vapor by polymer blends of crystalline and amorphous polymers in the range of small relative vapor pressures (up to the beginning of the isothermal melting) occurs in amorphous regions of the blend structure [15, 16]. In this context, the sorption data for all studied blends with a crystalline PEO were represented with allowances made for the proportion of amorphous regions in the system. In the calculation, we used the degrees of crystallinity that were determined for the same blends.

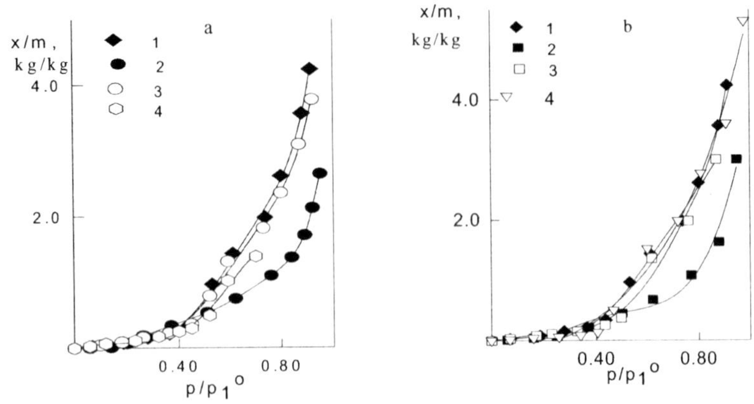

Figure 1. Sorption isotherm of chloroform vapor in the systems PEO-poly (2-vinyl pyridine) (a) and PEO-poly (4 -vinyl pyridine) (b): 1-PEO, 2 - amorphous P2VP or P4VP, 3 and 4 polymer blends with 10 and 25 mass.% of amorphous polymer respectively.

The amount of solvent (chloroform) sorbed by the amorphous part of the polymer mixture was calculated using simple relations following from the representation of the sample weight as the sum of weights of its amorphous m_a and crystalline m_{cryst} parts:

$$m = m_a + m_{cryst}, \qquad (1)$$

where $m_{cryst} = m\alpha$, and $m_a = m - m_{cryst} = m(1 - \alpha)$; α – is degree of crystallinity of the used blends. Hence, it follows that sorption of the solvent by amorphous part(x/m_a) of PEO and its blends is equal:

$$]x/m_a = x/m \bullet 1/(1-\alpha) \qquad (2)$$

were x/m is experimental sorption value of the sample.

All results are listed in the third part of Table 2.

Table 2. The Degree of Crystallinity of PEO Blends and the Gibbs Energy of Polymer Mixing.

System	PEO content of the mixture, wt.%	Crystallinity degree, α, %	Gibbs energy, $\Delta g_{x,amorph}$,kJ/kg
PEO-PMA	87	72	-4.8
PEO-PBA	83	65	-2.3
PEO-PMMA	85	60	-3.1
PEO-PBMA	80	70	-2.3
PEO-P2VP	90	64	-4.6
PEO-P4VP	90	56	-9.7

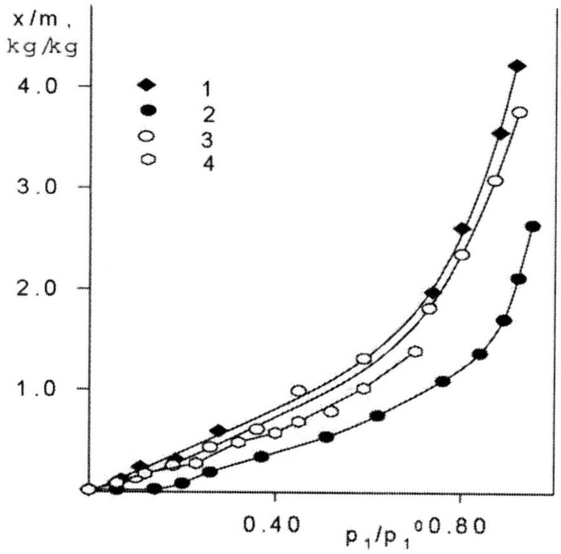

Figure 2. Sorption isotherms of chloroform vapor by polymer blends PEO - poly(2-vinyl pyridine) with allowances made for the contribution of amorphous regions: PEO (1), P2VP (2), PEO-P2VP (10%) (3) and PEO-P2VP(25%) (4).

Sorption isotherms of chloroform vapor by individual PEO and its blends with amorphous polymers, with allowances made for the degree of crystallinity of the systems, are presented for example in Figure 2 for the PEO and its blends with poly(2-vinyl pyridine).

It is shown that the sorption capacity of PEO-P2VP blends is lower than for the individual PEO. If the content of glassy polymer P2VP (or P4VP) is increased, the sorption capacity of chloroform vapor is decreased. Analogous regularities were received for the PEO-poly(acrylates) systems [13]. It was also received that the sorption capacity of PEO–PMMA blends with respect to chloroform is lower than that for PEO–PMA blends. The blends of PEO with methyl esters of poly(acrylic acid) and poly(methacrylic acid) showed a lower sorption capacity as compared to the corresponding systems containing butyl acrylates [13]. The peculiarities of the sorption behavior of PEO – poly(acrylate) and poly(methacrylate) blends can be associated with differences in the chemical structure of the amorphous component and its kinetic flexibility at the sorption experiment (see data in the Table 1), These factor are determined by the character of interaction between the tested polymer systems and the sorbate molecules. Therefore, the blends of PEO with amorphous polymers characterized by smaller molar volumes of their units (PMMA and PMA) show lower sorption capacities with respect to the solvent as compared to systems containing PBMA and PBA. The latter, owing to the presence of a longer aliphatic substituent, have looser molecular packing and higher kinetic chain flexibility. Under the test conditions, PMMA chains in the PEO – PMMA mixture, as soon as P2VP and P4VP in the mixture with PEO, occur in the glassy state which results in a lowered sorption capacity of the mixture as compared with the other systems.

From the sorption data obtained, we determined the specific Gibbs energy of mixing of all the studied systems with chloroform. The energy of mixing was determined by the chloroform vapor depression over polymers and their blends according to the procedure originally applied o polymer solutions in [6].

The experimental sorption data were used for the calculation of the specific Gibbs energy of mixing of all studied systems with chloroform. For this purpose the change in chemical potential of the solvent upon sorption $\Delta\mu_1$ was determined according to equation:

$$\Delta\mu_1 = RT/M_1 \ln p_1/p_1^0 \tag{3}$$

where M_1- molar mass of the solvent. The change in chemical potential of polymer and polymer blend $\Delta\mu_2$ was calculated by means of Gibbs - Duhem equation according to [17] and then, using these values, Gibbs energy of mixing of PEO and its blends with chloroform was determined:

$$\Delta g^m = \omega_1 \Delta\mu_1 + \omega_2 \Delta\mu_2 \tag{4}$$

where ω_1 and ω_2 are weight fractions of the components in chloroform solution. The concentration dependence of Δg^m are presented in Figure 3. The values of Δg^m are negative for all systems over the entire concentration range, the second derivative of Δg^m is positive. This means that all solutions of PEO and its blends in chloroform are thermodynamically compatible with solvent and homogenous.

A good thermodynamic affinity to chloroform (at minimal in the curves, Δg^m is in the range from –7 to –18-20 kJ/kg of the solution), is associated with a sufficiently strong molecular interaction of the polar carbonyl groups of poly(acrylates) with chloroform, which form C=O... H–CCl_3 hydrogen bonds. In the case of P2VP and P4VP H - bonds are formed by π-vacant orbital of heterocyclic ring and H-atom of chloroform. The hydrogen bonds of PEO - chloroform are formed between oxygen atom of the polymer chain and hydrogen atom of chloroform molecules. The latter agrees with the results of direct thermochemical measurements of the interaction energy of PEO with chloroform [18].

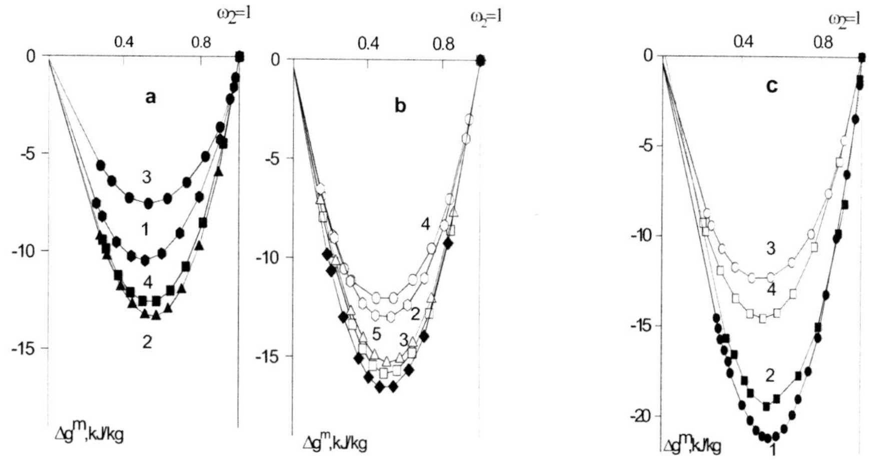

Figure 3. Concentration dependence of the average specific Gibbs energy of mixing with chloroform of: individual PMA(1), PBA(2), PMMA(3), PBMA(4) (a); PEO (1) and its blends with PMA(2), PBA (3), PMMA (4), PBMA (5) (b); P2VP (1), P4VP(2) and PEO-P2VP(3), PEO-P4VP (4) blends (c). Content of amorphous polymer in the blends is corresponded to the data of Table 2.

The data of Figure 3 were used for calculation the average specific Gibbs energy of mixing of PEO with amorphous polymers (Δg_x) at 25 ^{0}C. The calculation was carried out using the thermodynamic cycle described in detail in [5, 6], which involves the Gibbs energy of mixing of the components (ΔG_1 and ΔG_2) and their mixture (ΔG_3) with the same solvent (calculated per *g* of a polymer component):

$$-\Delta g_x = \Delta G_3 - (\Delta G_1\,\omega_1 + \Delta G_2\,\omega_2) \tag{5}$$

where ω_1 and ω_2 are the weight fractions of components in the polymer mixture. The values of ΔG_1, ΔG_2, ΔG_3 are equal to segment that is severed by tangent to the curves of $\Delta g^m = f(\omega_2)$ at $\omega_2 = 1$, that is presented in the Figure 3. These calculations were made for the contribution of amorphous looser paced regions to the mixing of the components. The result is listed in the last part of Table 2. The Δg_x values are negative and consistent with those obtained for a number of blends of amorphous polymers [4 - 6]. The negative Gibbs energy of mixing of PEO with the examined amorphous polymers indicates that the components are compatible in the amorphous phase. This result agrees too with the data of the parameters of the affinity of PEO with amorphous PMMA and PMA reported for the blends of approximately the same composition in [11, 19], based on the thermochemical data. The value of the Gibbs energy is depending on the structure of amorphous component. For the systems PEO-poly(acrylates) and poly(methacrylates) the Gibbs is more negative in the blends with methyl ethers (PMA,

PMMA) than butyl ethers (PBA, PBMA). Long aliphatic radical in the side groups of the molecules of amorphous component change for the worse interaction between polymers and value of Δg_x became less negative. For the systems PEO-poly (vinyl pyridines) the more negative value of Δg_x is in the mixture of PEO with P4VP, where active to interaction nitrogen atom is in Para-position to the carbon chain that is facilitate interaction between polymers.

Conclusion

It has been determined that the Gibbs energy of mixing of crystalline PEO with amorphous polymers poly(vinyl pyridines), poly(acrylates), that was calculated from data of the sorption of common for two polymers solvent, is negative and found to range from – 2.3 to – 9.7 kJ/kg. These data were received by using of the thermodynamic cycle for calculation of the Gibbs energy and taking into consideration that only amorphous part of PEO structure can interact with the second amorphous component of the blend. The results obtained in the present work agree with the notions that crystalline and amorphous polymers can be mixed only due to the penetration of the amorphous polymer chains in the amorphous regions of crystalline polymers [16, 20–22]. The allowance made for determination of the portion of amorphous structure in the polymer blends offers a means of correct estimation the thermodynamic affinity (the Gibbs energy of mixing) of the components; moreover, this allows better to reveal the effect of the structure of amorphous polymers on their compatibility with the crystalline polymer.

Acknowledgments

This publication was supported by Russian Scientific Foundation (grant 01-03-96454) and the Civilian Research and Development Foundation (grant REC-005)

References

[1] Krause S. in "*Polymer Blends*", Paul D.R., Newman S. Eds., New York: Academic, 1978.

[2] Paul D., Bucknall C. *"Polymer Blends. Formulation and Performance*". J.Wiley & Sons Ltd., London, 2000.

[3] Chalykh A.E., Gerasimov V.K., Mikhailov Yu.M. "*Phase Diagrams of Polymer Systems" (in Russ.*), Moscow: Yanus-K, 1998.

[4] Tager A.A., *Vysokomol. Soedin (in Russ.), A14:12* 2690 - 2706 (1972).

[5] Tager A.A., Sholokhovich T.I., Sharova I.M., Adamova L.V., Bessonov Yu.S., *Vysokomol. Soed. (in Russ., A17:12* 2766 - 2773 (1975).

[6] Tager A.A., Scholokhovich T.I., Bessonov Yu.S., *Eur. Polym. J., 11* 323 - 326 (1975).

[7] Wright P.V, *J. Macromol. Sci.-Chem., A26:2-3* 519 - 550 (1989).

[8] Armand M.B., *Solid State Ionics, 69:3* 309 - 319 (1994).

[9] Baril D., Michot C., Armand M.B., *Solid State Ionics, 94:*1 35 - 47 (1997).

[10] Morales E., Acosta J.L., *Solid State Ionics, 95:3-4* 269 - 274 (1997).

[11] Pedemonte E., Pollerri V., Turturro A., Cimmino S., Silvestre C., Martuselli E., *Polymer, 35:15* 3278 - 3781 (1994).

[12] Suvorova A.I., Hassanova A.H., Tyukova I.S., *Polym. Int., 49:9* 1014 - 1017 (2000).

[13] Suvorova A.I., Tyukova I.S., Khasanova A.Kh., *Polymer Science,A43:12* 1261 - 1265 (2001).

[14] Tager A.A., Tsilipotkina M.V., *Usp. Khim. (in Russ), 47:1* 152 - 175 (1978).

[15] Allen G., Booth C., Gee G., Jones M.N., *Polymer, 5:3* 367 - 375 (1964).

[16] Chalykh A.E., "*Diffusion in Polymer System"(in Russ.)*, Moscow: Khimia, 1987.

[17] Tager A.A., "*Physical Chemistry of Polymers" (in Russ),* Moscow: Mir 1978.

[18] Safronov A.P., Kovalev A.L., *Polymer Science, A41:6* 670-675 (1999).

[19] Pedemonte E., Burgisi G., *Polymer,365:15* 3719 – 3721 (1994).

[20] Chow T.S., *Macromolecules, 23:3* 333 – 337 (1990).

[21] Štarka J., Schmidt P., Dubal J., Schneider B., Spevačék B., *Polymer,36:9* 1147 – 1153 (1995).

[22] Parizel N., Laupretre F., Monnerie L., *Polymer, 38:17* 3719 – 3723 (1997).

In: Chemical Reactions in Liguid and Solid Phase
Editors: G. E. Zaikov and A. Jiménez, pp. 145-154
ISBN 1-59033-854-5

Chapter 7

THERMAL STABILITY OF NATIVE POLYBUTYLENE TEREPHTHALATE

V. N. Shelgaev, A. K. Mikitaev, S. M. Lomakin*, G. E. Zaikov* and E. V. Koverzanova**

GNU "Center on Composite Materials" of Ministry of Education of RF, 9 Miusskaya PI, 125047 Moscow, Russia
*N.M. Emanuel Institute of Biochemical Physics, HAS, 4 Kosygin Street, 119991 Moscow, Russia
** N.N. Semenov Institute of Chemical Physics, RAS, 4 Kosygin Street, 119991 Moscow, Russia

1. INTRODUCTION

As it was reported earlier [1], we have developed a new way to synthesize polybutylene terephthalate (PBTP) in the presence of stabilizing systems. The PBTP synthesized in such conditions had high physical and chemical characteristics including the enhanced thermal stability.

At the same time, it is known that the wide use of the native PBTP will be limited without careful physico-chemical investigation. Because of this, the study of the PBTP thermal and thermo-oxidative stability is the one of the priority tasks. The actuality of this problem is especially important when the material or composites based on it supposed to be used in the fields where the thermal stability is the main criterion. A special attention in this work was paid to the PBTP transformations in conditions close to the ones of the polymer processing.

We assumed that the understanding of the basic processes taking place at thermal and thermo-oxidative polymer destruction in these conditions will allow to develop more effective stabilizing systems for PBTP.

2. EXPERIMENTAL

As the object for investigation, we used the polybutylene terephthalate type 'MAKOOJI', synthesized in the presence of the catalytic system containing lrganox-1010, trinonylphenylphosphine and sodium hypophosphite taken in the ratio 0.3:0.1:0.05 mass % correspondingly. The polymer had the highest physico-chemical properties from the PBTP synthesized by us before [1].

The polymer samples were investigated both in the range of temperatures close to the PBTP processing temperatures (250°C-260°C) and at temperatures, at which the PBTP decomposition occurs with a significant rate (340°C-500°C).

The PBTP sample pyrolysis was conducted in the pipe pyrolytic cell under air flow (consumption 40 ml/min). The gaseous pyrolysis products were trapped by their passing through hexane cooled on the icy bath. The dry residue after evaporating of hexane was dissolved in tetrahydrofurane. The products were underwent esterification according to the procedure [2] to form the acid derivatives. After that, they were analyzed.

The screening analysis of the pyrolysis products was conducted by using the gas chromatograph 'LjBer-500M' equipped by the electron trap detector. To separate the destruction products, we used the glass filled column length 4 m with an inner diameter 3 mm. The stable phase was phenylmethylsilicon OV-17; the column and ejector temperatures were 230 °C, the detector temperature was 290 °C, the gas-carrier was nitrogen, the consumption - 25 ml/min. The probe volume was 2.0 ul.

The complex chromato-mass-spectrometric analysis was conducted by using the gas chromatograph ' Varian 3300' with the mass-spectrometric detector type the 'ionic trap' 'Finnigan VAT ITD 800'. The product separation was conducted on the quartz capillary column length 30 m, inner diameter 0.32 mm with the stable phase DB-5 (phenylmethylsilicon, film thickness 0.25 p,m). In the course of analysis, the column temperature was risen from 100 °C to 270 °C with the rate 10 °C/min. The injector temperature was 200 °C. As the gas-carrier, helium was used. The gas pressure at the entrance was 0.1 MPa. The probe in amount of 1.0 ul was incorporated into the chromatograph (without the stream division). The time to the probe incorporation was 30 sec. The mass-spectra were registered at ionization by the electron strike with the energy 70 eV, the scanning rate was one mass-spectrum/min, the scanning diapason was 40-650 a.e.m.

3. RESULTS AND DISCUSSION

The data in Fig. 1 show the yield of some oxygen-containing gaseous products forming in the process of the PBTP thermo-oxidative destruction. The data in Fig. 1 also show the yield of the analogous products formed in the process of PBTP destruction in the helium atmosphere. As it is seen, in the presence of oxygen, the yield of all destruction products is significantly increased. Besides, the carbon monoxide is formed, which was absent at PBTP destruction in helium.

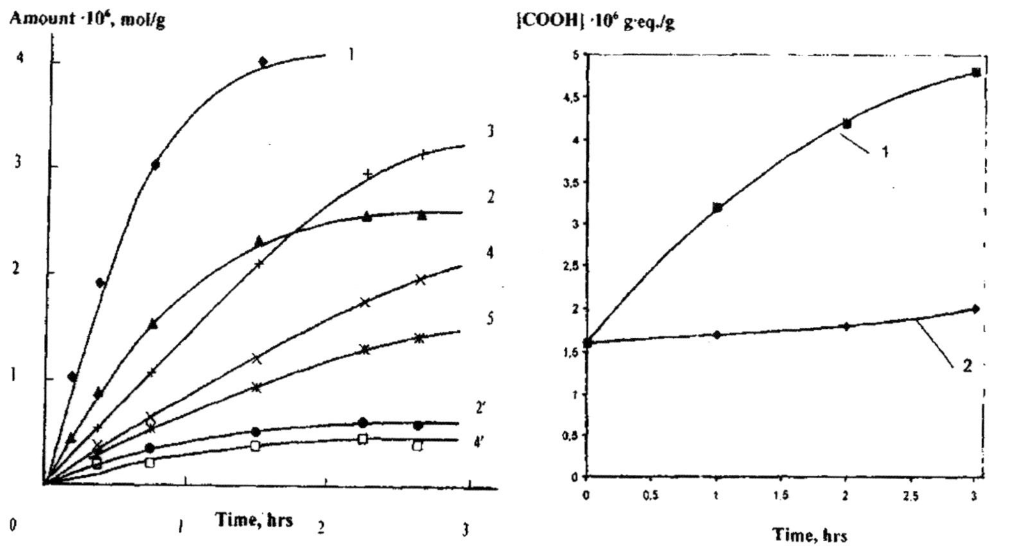

Figure 1. Kinetics of oxygen absorption (curve 1) and gaseous products separation at thermo-oxidation (curves 2-5) and the Figure 2. Change of the COOH-groups amount with thermal destruction of polybutylene terephthalate (curves 2' and time at heating of polybutylene terphthalate in 4') in helium. 1.-O_2; 2,2'-CO; 3.-CO,; 4,4' -H,O; 5.-CH_3O. P_{02} different gaseous media (270°C, 3 hrs): 1.-in oxygen = 40kPa; T= 270°C. P_0,=40 kPa; 2.-in vacuum.

The noticeable destruction rate increase in presence of oxygen can be observed also on the fast accumulation of carboxyl groups in the polymer (Fig.2, curve 1). It follows from the obtained data that the thermo-oxidative destruction of PBTP, probably, occurs through the stage of hydroperoxides formation. This is confirmed by the water yield increase in presence of oxygen. The absorption of a significant amount of oxygen (Fig.l, curve 1) at the PBTP oxidation witnesses the oxygen participation not only in the initiation but also in the further stages of the radical chain process.

The thermo-oxidative destruction of PBTP occurs, probably, in accordance with the same mechanism as the liquid-phase oxidation of the low molecular hydrocarbons [3]. The initiation of the radical chain process starts as the result of interaction of oxygen atoms with the atoms of hydrogen of the polymer chain methylene fragments, formation of hydroperoxide, its decomposition, and the consecutive fragmentation of the polymer chain. This is accompanied by separation of the gaseous and

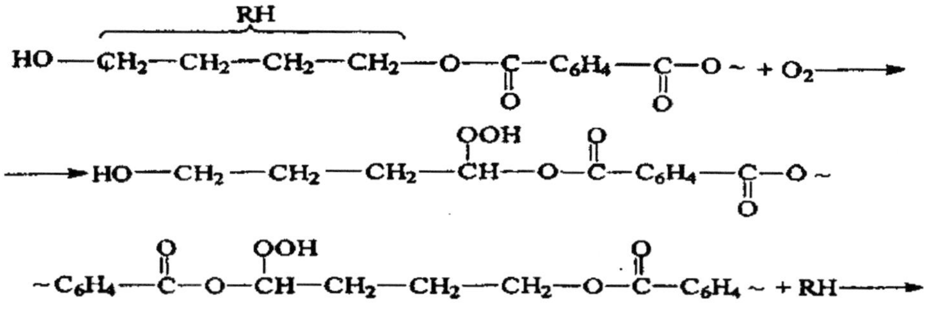

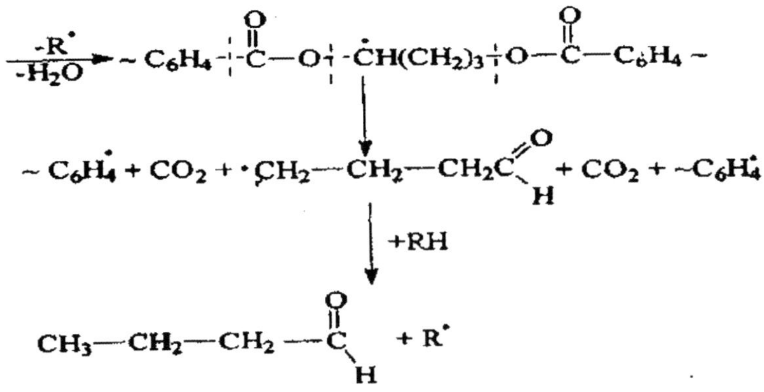

It is probable that along with the hydroperoxide formation, the isomerization of the peroxide radical with a consecutive decomposition on the bond -O-O- takes place. This will lead to formation of aldehyde [4]:

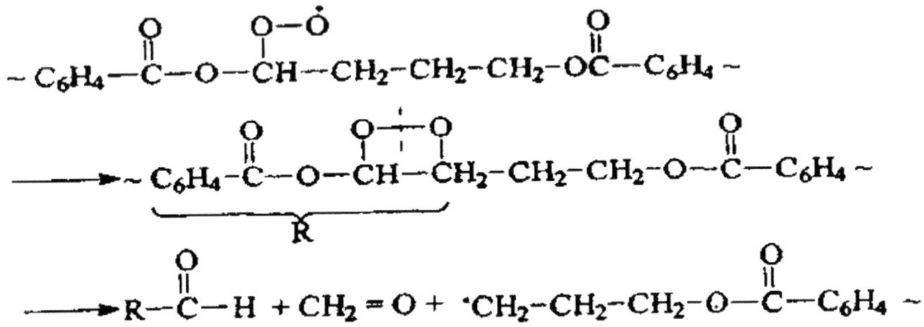

As the PBTP thermo-oxidative destruction proceeds, the mechanism of the process becomes more complicated, the variability of the active centers and the intermediate and the end products increases.

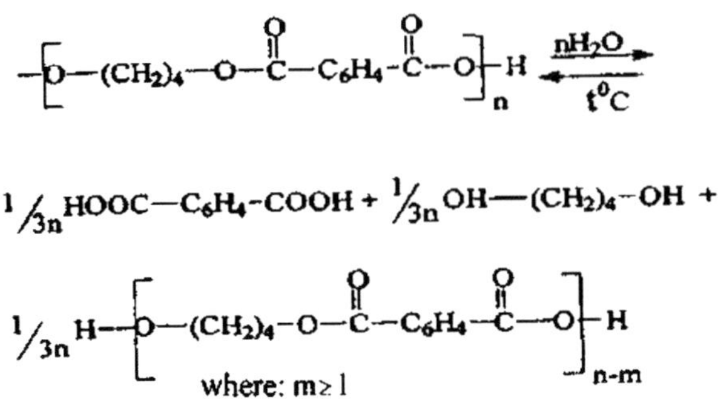

Along with the radical chain processes, the hydrolytic processes take place. They are accelerated by the formed acidic decomposition products in the course of the thermo-oxidation process (see scheme on the left). The formation of various acidic products can occur also at the hydroperoxide decomposition as in the case published in [5] (see scheme below)

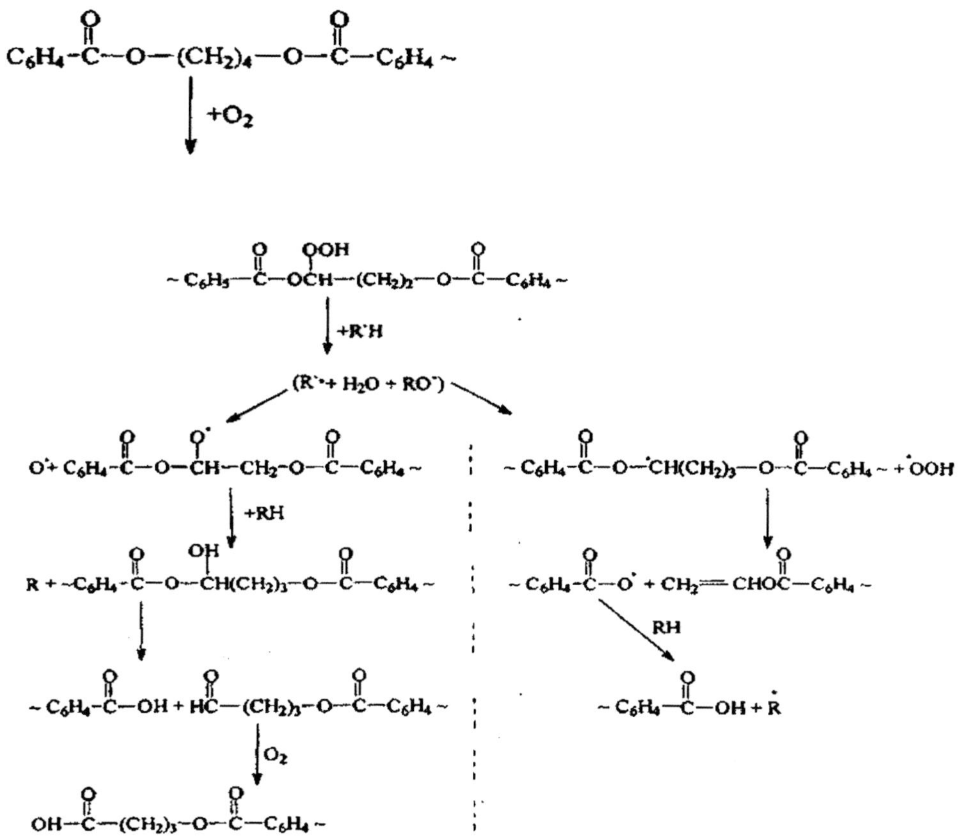

Thus, in the process of PBTP thermo-oxidative destruction at moderate temperatures the following products are formed water, aldehydes, and others. The products at elevated temperatures at synthesis and in the melt on the stages of processing can cause the hydrolysis of the ester bonds what will negatively affect on the molecular mass value, and, as the result, on the polymer properties. This was the reason to conduct PBTP synthesis in the presence of the stabilizing systems.

In the case, when the PBTP thermo-oxidative destruction occurs at the enhanced pyrolysis temperatures, the composition of the decomposition products and their amount is noticeable changing. The data in Table 1 show the composition of the PBTP destruction products formed at thermo-oxidation at 340°C and 520°C. At these temperatures, there is a mass PBTP decomposition according to the data of the thermogravimetric analysis.

Table 1. Products of the PBTP Thermo-Oxidative Destruction at Temperatures of 340°C and 520°C

Compound name	Separation time, mm.	Hexane extract		Products dissolving inTHF		Methyl derivatives	
		Peak Intensity	%	Peak Intensity	%	Peak intensity	%
Methyl ether of terephthalic acid	6:23	_	_	.	_	1227	0.5
Benzaldehyde	7:11	933	4.6	1875	0.5	163	O.I
4-[(Butenyloxy)carbonyl]-benzoic acid	8:43	-	-	1465	0.4	207	0.1
Dimethyl ether of terephthalic acid	9:04	-	-	1152	0.3	134571	51.1
4-Hydroxybutyl-4-formyl benzoate	11:53	1119	5.1	2858	0.8	819	0.3
Methyl-4-(4-pentanoyl) benzoate	12:17	-	-	-	-	109687	41.6
1 -Phenyl- 1 -pentanone	12:45	715	3.6	3910	1.0	1230	0.5
Terephthalaldehyde acid	12:59	-	-	4501	1.2	-	-
4-Hydroxybutyl-4-acetyl benzoate	13:11	2033	10.1	14266	3.7	1925	0.7
1 -benzocarboxy-4-methoxy- 1 -benzocarboxy anhydride	13:57	418	2.1	2854	0.8	-	-
3-Butenyl-4-formyl benzoate	15:13	3930	19.6	147510	38.8	1907	0.7
1 - Benzocarboxy-4-pentanoy I- 1 benzocarboxy anhydride	15:19	1005	5.0	11730	3.1	-	-
4-Benzocarboxy-4-[(3-butenyloxy)-carbonyl]- 1 -benzenecarboxy anhydride	15:50	4417	22.0	86666	22.8	4371	1.7
1 -(4-Hydroxybutyl)-4-methyl terephthalate	16:04	2136	10.6	45576	12.0	2362	0.9
Non-identified compounds	-	3493	17.4	56118	14.7	3786	1.5
		20099	100	380481	100	263374	100

Table 1. Products of the PBTP Thermo-Oxidative Destruction at Temperatures of 340°C and 520°C (continued)

520°C							
l-(3-Butenyl) benzene	5:05	.	_	47457	17.3	.	.
Benzaldehyde	7:13	6886	63.6	14465	5.3	289	0.5
Dimethyl ether of terephthalic acid	9:04	-	-	-	-	39491	63.6
Diphenyl ketone	10:49	-	-	1209	0.4	135	0.2
4-Hydroxybutyl-4-formyl benzoate	11:56	213	2.0	3330	1.2	262	0.4
Methyl 4-(4-pentanoyl) benzoate	12:18	-	-	-	-	13326	21.5
1 -Phenyl- 1 -pentanone	12:45	168	1.6	3441	1.3	481	0.8
Terephthalaldehyde acid	12:55	213	2.0	3330	1.2	113	0.2
4-Hydroxybutyl 4-acetyl-benzoate	13:11	420	3.9	15086	5.5	745	1.2
1 -Benzocarboxy-4-methoxy- 1 - benzocarboxy anhydride	13:58	-	-	2289	0.8	146	0.2
3-Butenyl 4-formyl benzoate	15:15	-	-	33398	12.2	436	0.7
1 -Benzocarboxy-4-pentanoyl- 1 benzocarboxy anhydride	15:21	-	-	6031	2.2	238	0.4
4-Benzocarboxy-4-[(3-butenyloxy)-carbony 1]- 1 - benzenecarboxy anhydride	15:50	716	6.6	78092	28.5	2104	3.4
1 -(4-Hydroxybutyl)-4-methyl terephthalate	16:04	500	4.6	22182	8.1	966	1.6
non-identified compounds	-	1708	15.8	43558	16.0	3390	5.6
		10824	100	273898	100	62122	100

As it is seen from the data, at the PBTP thermo-oxidative destruction at 340°C, the basic products are the terephthalic acid the 4-4(pentanoyl) benzoic acid, and also 4-hydroxybutyl-4-acetyl benzoate, 3-butenyl-4-formyl benzoate, l-(4-hydroxybutyl)-4-methyl terephthalate, and 4-benzocarboxy-4-[(3-butenyloxy) carbonyl]-1-benzocarboxy anhydride. At increasing the pyrolysis temperature to 520°C, two new compounds appear: benzaldehyde and l-(S-butenyl) benzene. At the same time, the concentration of 4-hydroxybutyl-acetyl benzoate and 3-butenyl-4-formyl benzoate decreases approximately in three times. The concentration of l-(4-hydroxybutyl)-4-methyl terephthalate decreases in 1.5-2.0 times. The ratio between 4-(-pentanoyl) benzoic and terephthalic acids was also changed.

It is possible that in condition of the heat wave in air, the simultaneously occurring processes take place: the radical chain reactions of oxidation and the occasional homolytic decomposition of polymer chains. From the position, it is possible to explain the existence of such wide range of various products.

We have also investigated the kinetics of the thermal PBTP destruction at relatively low temperatures close to the processing temperatures. By using the method of the pyrolytic mass-spectrometry, it was shown that thermal destruction of PBTP in the temperature range 250-350°C is accompanied by formation of carbon oxides, water, hydrocarbons (predominantly butadiene), benzoic and terephthalic acids. Besides, the mixture of oligomeric ethers is also formed.

$$-\overset{\displaystyle O}{\overset{\|}{C}}-Ar-\overset{\displaystyle O}{\overset{\|}{C}}-(CH_2)_4-O\sim \longrightarrow$$

$$\longrightarrow \sim C-Ar-C{\equiv}O^{+} + {}^{\bullet}O-(CH_2)_4-O$$

The appearance in the mass-spectra ions with high molecular mass at the PBTP pyrolysis, is, probably, related, as in case of PET [6], with the rupture of the 8-bonds in the ester fragment of the macrochain:

$$HO-\left[C(O)-Ar-C(O)-O(CH_2)_4-O\right]_m-C(O)-Ar-C{\equiv}O^{+}$$

In the mass-spectrum, the intense peaks of ions were found:

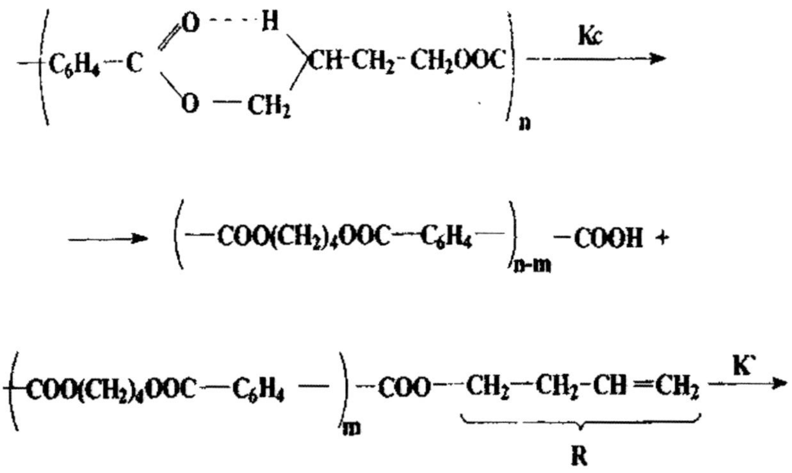

The ions are formed at the destruction products ionization. The destruction of the carboxyl-containing products of decomposition leads to appearance of carbon dioxide in the gaseous phase.

The data in Fig.2 and the kinetic curves prepared from the experimental data show the character of the change of the number average state of polymerization and the change of the end carboxyl groups content at different temperatures. The linear character of the values dependency witnesses that the PBTP destruction occurs randomly. It is possible that, as in the case of polyethylene terephthalate, the PBTP thermal decomposition on the stage of initiation is originally accompanied by formation of the cyclic transitional state in an occasional spot of the polymer chain with its consecutive decomposition [7] (see scheme above).

At higher temperatures, the destruction of butene-ether end groups is observed (see scheme below).

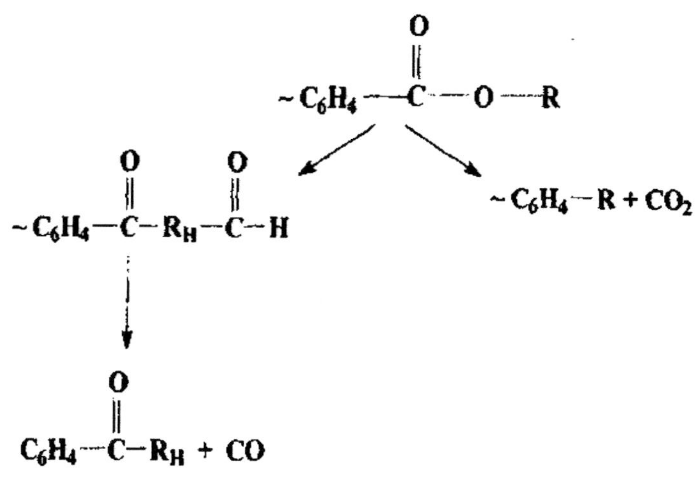

One of the ways of the PBTP thermal decomposition can be the rupture of the ether bonds with formation of the hydroxyl group caused by the migration of the methylene group hydrogen atom.

$$^{\bullet}O\text{-}(CH_2)_3\text{-}CH_2\text{-}O(O)C\text{-}C_6H_4\text{-}C(O)O\text{-}(CH_2)_4\text{-}O{\sim} \rightarrow HO\text{-}(CH_2)_4\text{-}O$$

This way of the reaction, probably, occurs at enhanced temperatures (above 300°C): This is supported by the significant amount of the hydroxyl-containing compounds formed at high temperature PBTP pyrolysis (Table 1).

CONCLUSIONS

The thermal and thermo-oxidative stability of polybutylene terephthalate (PBTP), synthesized in the presence of the optimal stabilizing system, has been investigated. It was determined the basic products of the PBTP destruction and the regularities of their formation in the temperature regime close to the temperature regimes of its processing.

The thermo-oxidative PBTP destruction has the radical-chain character. It starts from the oxidative decomposition of the aliphatic part of polymer. Parallel to the process, there are thermo- hydrolytic processes. The thermal destruction of polybutylene terephthalate occurs randomly. The conducted investigations allowed to make recommendations on the temperature regimes of PBTP processing because even a small excess of the chosen temperatures (~265°C) leads to the polymer chains fragmentation.

REFERENCES

[1] Mikitaev A.K., Storozhuk 1.P, Shelgaev, Alakaeva Z. T. // *Plast. Massy* (Russ.), 2001.

[2] Lurie A.a. *'Chromatographic materials'*(Russ.), M. Khimiya, 1978.

[3] Polymer Encyclopedia (Russ.), v.3, M: *Sovelskaya Enzyklopedia*, 1977, pp.624-626.

[4] Kovarskaya B.M., Blumenfeld A.B., Levantovskaya 1.1. *'Thermal stability of heterocyclic polymers'* (Russ.) M:, Khimiya, 1977, p. 74.

[5] Buxbaum LN. //*Am. Chem. Soc., Polymer Prep.*, 1967, v.8, pp.552-557.

[6] LudenvaldJ., Urrutia H. //*Macromol. Che.,* v.177, S.2079.

[7] Passalama V. // *Polymer*, 1976, v.17, p. 1044.

In: Chemical Reactions in Liguid and Solid Phase
Editors: G. E. Zaikov and A. Jiménez, pp. 155-163
ISBN 1-59033-854-5

Chapter 8

THERMAL OXIDATION OF SELF-DEGRADABLE COMPOSITE FILMS BASED ON LDPE

A. A. Ol'khov, L. S. Shibryaeva*, A. L. Iordanskii** and G. E. Zaikov*

M. V. Lomonosov Academy of Fine Chemical Technology,
86 Vernadski Pr., Moscow 117571, Russia
*N.M. Emanuel Institute of Biochemical Physics,
RAS, 4 Kosygin Str., Moscow 117334, Russia
**N.N. Semenov Institute of Chemical Physics,
RAS, 4 Kosygin Str., Moscow 117334, Russia

ABSTRACT

The effect of the structure of the composite extrusion films based on the mechanical blend of low density polyethylene (LDPE) and polyhydroxybutyrate(POB) on the set of service characteristics and the kinetics of the thermal oxidative destruction has been investigated. It was found that the aggregate state of polymers affects the value of the boundary surface in the blend films. The increase of the latter affects the changes of the conformation state of both polymers in the blends. In this case, the strength decreases, the steam permeability increases, and the processes of the thermo-oxidative destruction of the PE matrix on the beginning stages are accelerated.

1. INTRODUCTION

At the present time, the composite materials based on the traditional synthetic polymers and new biodegradable polymers have a big practical interest. As the destructible components, different biopolymers can be used, for example, polylactones [1], and polyorthoesters [2]. Among the latter, polyalkanoates [3] should be mentioned, and, in particular, their typical representative - poly-R-(3-oxybutyrate) (FOB). The composites of a such type belong, as a rule, to the thermodynamically incompatible blends. Such blends are characterized by the polydisperse structure with a developed boundary surface between phases. The polydispersity of the compositions, and, hence, the length of the interphase boundary in the blend depends on the nature of the contacting polymers [4], their rheology [5], cohesion forces, which characterize the intermolecular interaction [6], and also, on the mixing technology (rate, stress, temperature) [7]. At the blend forming, the components' structure undergoes significant changes on the molecular, supra-molecular, crystalline, and macro- levels compared with the original polymers. In its turn, the component structure determines both the value and the geometry of the boundary surface (BS) and the depth of the boundary layer. The BS parameters affect the service properties, kinetics of the diffusion and chemical processes, in particular, the processes of the composite oxidation [4, pp. 494-536].

Taking into account the important role of the chemical stability toward the air oxygen action, the goal of this work was to establish the relationship between the morphology of the heterogenic blend films based on the blend LDPE and FOB and their kinetics of oxidation.

2. Experimental

The low density polyethylene type FISHO 15803-020, having M, = $2.5\text{-}10^5$, as granules size 2-3 mm and powder particle size 5-10 urn, and FOB (BIOMER, Germany), having M_n = $2.0\text{-}10^5$, as powder with the particle size 7-12 μm were used.

The composition PE/POB (92/8 mass % ratio) was prepared in the following way. By using the ball mill, the granulated PE was mixed mechanically with the FOB powder and the powdered PE was mixed with the powdered FOB. After that, by using the one-screw laboratory granulator (L/D = 20, D=20 mm, V=100 rev/min, T=l 25 °C), the blend granules were obtained. Then, the granules were put into aggregate for making the 'hose' type film APri-20 (D=20 mm, L/D =25, V=100 rev/min, temperature along the extruder zones was less than 185 °C) and the film 'hose' was formed. The thickness of the film along the perimeter was 45±5 urn at the drawing ratio of 5, and the state of expansion by air was 2.

The FOB state of dispersity in the samples was determined by the method of the raster electron microscopy on the longitudinal the cross (toward the extrusion direction) crushed sample surfaces at magnitude 100-10,000. The used electron microscope was "BS-301" (TESLA, USA) type. The heat-physical parameters of the blends were determined by using the differential scanning calorimeter DCM-2 at the scanning rate 16 Deg./min. The calibration of the temperature scale was conducted based on indium T_{melt}=l 56.5 °C, the sample weight was 14-15 mg. The structure and the state of orientation of the polymer chains were determined by using the FT-IR spectroscopy method, the device was "Bruker IFS-48" (Germany). For this purpose, we used the following structure-sensitive absorbency bands: 729 cm^{-1} (the band connected with PE crystallinity [8]), and 1228 cm^{-1} (the band connected with the FOB crystallinity [9]). The spectra were registered in the polarized light. The beam was polarized along and perpendicular to the extrusion direction. The optical densities of each band were determined in accordance with the formula

$$D_0=D_H + 2D\wedge$$

where: D_u is the optical density along the extrusion direction; D^-perpendicular to the extrusion direction. The molecules orientation was evaluated based on the dichroic ratio R= D_H / $D_\pm$ of the structure-sensitive bands.

The strength properties of the films were determined according to COST 14236-81. The water penetration at 18 °C was determined according to the procedure [10]. The oxygen absorption was measured by using the manometric device [11].

3. Results and Discussion

The microscopic investigations of the PE/POB blends showed that the FOB phase has clearly defined boundaries. This witnesses the absence of the noticeable transition interphase layer. Fig. 1 shows the size distribution curves of the FOB phase in PE. Based on the obtained microphotographs and by using the Kavaliery-Akker [7, pp. 136-139] principle, the average sizes of the FOB phase were calculated. After that, the differential distribution curves were built. It is seen from the data that the average size of FOB in the films, formed from the system 'granules-powder' is about 1.0 μm. In the films prepared from the system 'powder-powder', the average size of the FOB is 0.3-0.6 μm. Because the concentration of FOB in the compositions is the same, the films prepared by blending the two powders have the more developed specific boundary surface. Due to the fact that we did not observe the dependence on the initial aggregate state of the boundary surface between the two phases, blending LDPE and FOB particles: a/ LDPE In our case, the determining value will have the particle size granules + FOB powder; b/ LDPE powder + of the dispersion phase and the length of the boundary POB powder between phases. This is confirmed by the literature data [12; 5, pp.276-277] where the authors demonstrated the relationship between the initial particle size of the dispersion phase and the final dispersity of the obtained composites, which affects the service properties level. The more dispersed structure forms at blending of the same particle size components. In case the system components have quite different particle sizes, their behavior at blending, will be different under action of the gravitational forces on the particles having significantly varying masses.

From the published data, it is known that BS can significantly affect the strength properties [7, pp.230-235] and the compound transport characteristics [13] of the polymeric materials. To evaluate the role of the BS on the structure and properties of the films based on LDPE/POB, we have determined the structure parameters (crystallinity and orientation state of the crystalline phase macromolecules) and the service properties (strength and the water penetration) of the films (see Table 1).

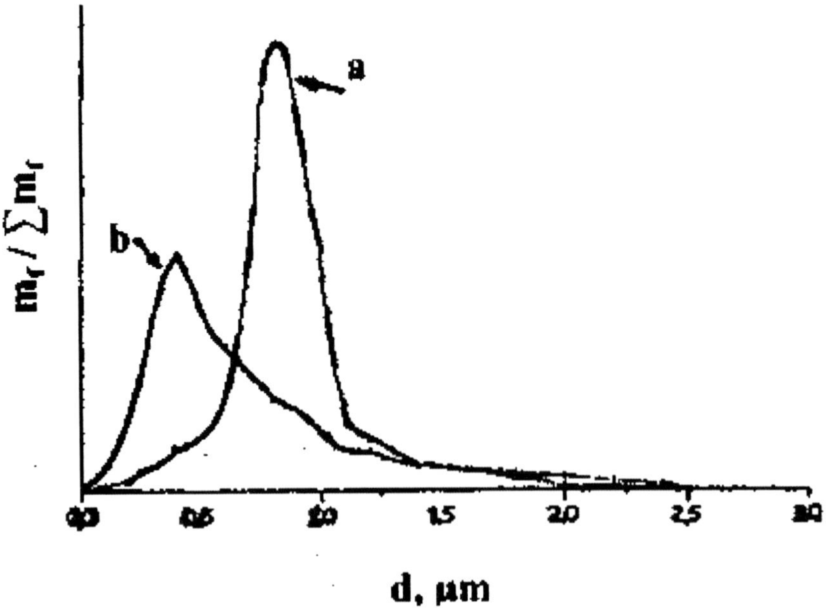

Figure 1. Size distribution curves of FOB developed interphase layer

From the data in Table 1, it is seen that the composite films have decreased vales of tensile strength and increased values of the water penetration compared with the monolith PE films. The least strength and the highest water penetration were found in the films with the larger boundary surface. This is explained by the interphase layer absence. In this case, the structure of polymers in the boundary proximity is less dense compared with the one in the volume. Besides, the components by themselves can form microscopic channels along the phase boundary, which can occur in the course of film drawing and expansion of the film hose. It is known from the literature that the supra-molecular structures and the molecular orientation in polymer blends and the character of the components interaction [11, p.252; 14] affect significantly the oxidative processes. Because of this, in this work, we have conducted investigations of the crystalline and the molecular structures of the LDPE/POB blends.

From the data in Table 1, it is seen that melting temperature of the components does not practically depend on the BS value (the state of dispersity), and does not differ from the characteristic values of the films obtained from the original homopolymers. That confirms once more that in the blend films there is no intermolecular interaction between LDPE and POB, which would lead to formation of the clearly expressed interphase layer.

The state of crystallinity of the blend components decreases compared with its value for the original homopolymers. This decrease is, probably, related, in the first place, with the intermolecular interaction of components on the interphase boundary, i.e., with the effect of the BS on the processes of the structure formation.

Table 1. Structural and Service Parameters of Films Based on LDPE and FOB

System type	Melting temperature, °C	Crystallinity state, %	Tensile strength[A], MPa	Water permeability at!8°C,-10^8 g-cm/cm^2-hr-mm Hg	Optical Density, Rel. Units		Dichroism, Rel. Units	
					Band 729 cm'''	Band 1228cm-'	Band 729 cm'[1]	Band 1228cm'''
LDPE granules + POB powder	105.0*	21.8	18.0	5.52	1.460	1.420	0.939	1.400
	171.2**	31.4						
LDPE powder+ POB powder	106.0[AA]	19.1	15.6	11.6	1.420	1.520	1.070	0.720
	172.0	31.4						
LDPE	107.0	35.0	20.0	2.38	1.500	-	1.750	-
POB	175.4	70.0	40.0	0.013	-	5.860	-	2.500

*For LDPE granules only. **For POB powder only. [A]Along the extrusion direction. [AA]For LDPE powder only.

The change of the chain conformation (Table 1) also affects the processes of structure formation. Because of this, we used the data of the polarized IR-spectroscopy to investigate the features of the molecular orientation of components in the blends. The values of the optical density of the crystallinity blends are well correlated with the DSC data and confirm the amorphization of the polymer crystalline phases in blends compared with the homopolymers (Table 1). The existence of molecular formations orientation, determined by the dichroism of the absorbency bands, indicates on the blend component interaction at the interphase boundary, and it can be used for the qualitative evaluations. It is also seen from Table 1 that the state of the PE orientation in the polymer blend films decreases compared with the films of the original PE, but it practically does not depend on the BS value. The change of the crystalline regions orientation value in the PE matrix is, probably, due to the change of the relaxation processes rate caused by the intermolecular interaction in the forming boundary layer (adhesion layer) in the blends with POB. From the data of Table 1, it is seen also that the state and direction of the orientation of the POB crystalline regions of the phase in PE significantly depend on the value of the BS, and they also differ from the parameters of the original POB film. It can be concluded that, mainly, the change of the spatial orientation of macromolecules in the POB crystallites and the crystallizing processes rate is the function of the state of the composite dispersity. In other words, the smaller the disperse phase particle average size the bigger effect has the value of the boundary surface and connected with it the adhesion layer value on the disperse phase structure. The following regularity support the statement: the average particle size of the POB phase decrease in two times corresponds to the difference in the orientation states of the POB crystalline phase also in two times.

By comparison the orientation state of PE and POB in the blend films with different area of the boundary surface, it is possible to note the discrepancy of the numeric values of the orientation parameters (R) of these phases, which is, probably, connected with the different rate of the relaxation processes occurring in the PE matrix and in the POB phase at the film formation. That can indirectly witness the absence of the transition layer (interphase layer) in blends, i.e., the absence of the intermolecular interaction of the PE phase with the POB phase in the compositions.

The macromolecules located directly in the vicinity of the phases boundary have less mobility compared with the molecules in the polymer volume. The effect of the boundary surface can be transformed into the polymer phase volume caused by the intermolecular interaction [5, pp.158-164]. The effect of the boundary surface can be also revealed in the retardation of the relaxation and crystallization processes [14, pp.30-33]. As it was mentioned before, we could not find existence of the transition layer in the investigating blends. Because of this, we can assume existence of only thin adsorption layer thickness less than 5-10 nm [15]. This layer does not show the properties of a separate phase due to its small thickness compared with the thickness of the boundary layer adjacent to the phase boundary.

According to the theory of the adhesion interaction [5, pp. 154-186], on the boundary surface, macromolecules are in the oriented state. The cohesion strength in this case increases causing the change of the chain conformation and, hence, the macromolecule segments freedom states decrease. As it is seen from Table 1, the PE crystallinity state in the blends practically does not change in dependency on the BS value. It is possible that the PE matrix has a low sensitivity toward the change of crystallinity in the BS due to a small content of POB. On the contrary, the state of POB phase crystallinity is changing at the BS value change. In the blends with higher value of BS, the POB phase has higher crystallinity. This effect can be explained by the different course of the structure formation processes on the molecular level at the polymer melt flow and film samples formation.

Based on the discussion above, it can be assumed that the state of dispersion of the biodegradable polymer POB in the LDPE matrix will affect the kinetics of the oxidation process of the composite film materials based on their blend, what we observed from the data shown in Fig.2.

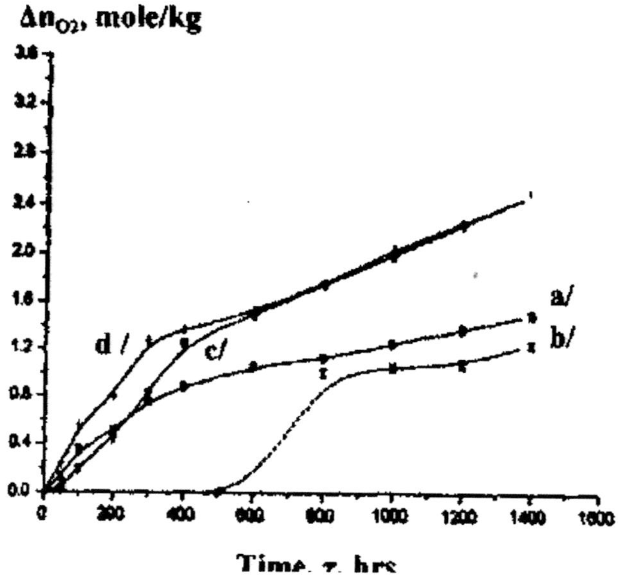

Figure 2. Kinetic curves of oxygen absorption on the polymer layer adjacent to the boundary surface.

The data in Fig.2 show the kinetic curves of oxygen absorption by samples of PE, POB, and their blends. As it is seen, the kinetics of the blend samples oxidation differs significantly from the kinetics of the homopolymer samples oxidation, and also the kinetics are deviated from each other. At deep states of oxidation (more than 0.5 mole/g), the sample with a higher BS is oxidized practically with the same rate as the sample with a lower BS, but both of them are oxidized significantly faster compared with the original PE.

At the beginning, oxidation step (below 0.5 mole/g), the samples with a higher BS value are oxidized faster compared with the original PE sample and the sample with a lower BS, and the original POB sample does not oxidize at all in the range of temperatures. This is confirmed by the value of the oxidation induction period. The last one is accepted to be the time till the oxidation process branching (when the process becomes auto-accelerated) and the change of the oxygen absorption rate in time.

By comparison the induction period and the rate of the oxygen absorption with the value of the boundary surface, it is seen that the sample with the more developed BS has the highest oxidation rate. This is connected, firstly, with the fact that FOB at elevated temperatures (processing and the thermal oxidation) undergoes destruction and, therefore, is the potential source of the low molecular radicals, which can initiate the oxidation in the PE matrix [16] caused by the phases contact on the boundary and the free valency transformation. This will initiate the oxidation process in the composite volume.

Secondly, the higher the specific boundary surface the larger is the contact area of the FOB phase with the PE matrix. As it is seen from the data in Table, the orientation state of the PE is larger compared with the smaller BS. The PE macromolecules state of orientation increase leads to the oxidation resistance decrease [17]. This can explain the high oxidation rate in case of a larger boundary surface.

As it was shown, the character of the PE/POB blend structure parameters change on the molecular level is such that it should be expected the reduction of the polymer compositions reaction ability. It is known that the segmental mobility decrease, i.e. the enrichment of the polymer chains by the straightened conformations, retards the oxidative processes [18]. In our case, it is observed an opposite effect. It is connected, as it was mentioned before, with the increase of the polymer chains stress in proximity of the boundary surface. At the same time, the FOB can suffer the thermal oxidation destruction (at processing) and can be the source of the low molecular radicals, and, hence, FOB can be the initiation factor. In other words, it can initiate the PE oxidation reactions. In the blends of different polymers, such effects were observed in the work [19]. It should be noted that the effect depends on the BS value.

As it is seen from Fig.2, at low oxidation states (at the beginning stage of oxidation), the process kinetics is different for the blends. In the blends having a low BS value, the rate of oxidation is lower. At the same time, the original FOB does not oxidize. Such, not similar behavior of blends at oxidation is the sequence of the conformational changes in the PE matrix on the molecular level under action of the FOB phase. Apparently, at deep oxidation stages (the oxidation time is more than 700 hrs), FOB begins to oxidize forming the low molecular radicals, which accelerate the polymeric matrix oxidation process. This is supported by the jump on the FOB oxidation curve. In this case, the kinetic curves of the blends are coming close to each other and become more linear. This can witness the independence of the oxidation process of the blends on the state of dispersity (BS) because the amount of radicals forming by the FOB phase in the both blends are the same due to the same content of FOB in the compositions.

4. Conclusions

Based on the obtained results, we can conclude that:

- The aggregate state of the original polymers affects the value of the boundary surface (BS). The BS increase in the films based on the low density polyethylene (LDPE)/polyhydroxybutyrate (FOB) blend reduces the physical properties and significantly increases the transport characteristics.
- The state of the FOB crystalline phase depends on the value of BS: the more the value of BS the higher the crystallinity state.
- The state of dispersity of the FOB is assumed to affect the processes of the molecular orientation of the crystalline phase.

Thus, the BS increase changes the conformational state of polymers in the blends on different levels of the structural organization. This accelerates the effects on the thermal destruction at the beginning stage. This regularity can be expected also in case of the use of the secondary LDPE.

REFERENCES

[1] Aub'm M., andPrud'Hommei/*J. Polym. Science: Polym, Phys. Ed.*, 1981, v. 19, No.8,pp. 1245-1253.

[2] Fayt R., Jerome R., Teyssie Ph. //*J. Polym. Science: Polym. Phys. Ed*, 1981, v.19, No.8, pp. 1269- 1272.

[3] Babel W., Riis V., Hainich E.//*Plaste und Kautschuk*, 1990, B.37, No.4, S.109-115.

[4] *'Polymer Blends'*, Eds. D. Paul and S. Newman (Russ.), M. Mir, v. 1, 1981, pp.26-127.

[5] Lipatov Yu.S., *'Colloid chemistry of polymers'* (Russ.) Kiev, Naukova Dumka. 1984, pp.258-260.

[6] Povstugar V.I., Kodonov V.I., Mikhailova S.S. *'Structure and properties of the polymeric materials surface'* (Russ.), M. Khimiya. 1988, pp. 75-77.

[7] Kuleznev V.N. *'Polymer blends'* (Russ.), M. Khimiya, 1980, pp.94-106, 136-146.

[8] Elliot A. *'IR-spectra and polymer structure'* (Russ.).M. Mir, 1972, p.82.

[9] Lambeck G., Vorenkamp E.J., Schouten A.J. //*Macromolecules*, 1995, v.28, No.6. pp.2023-2032.

[10] *'Practice book on polymeric materials'* Ed. P.G. Babaevskii, (Russ.) M. Khimiya, 1980, pp.46-47.

[11] Shlyapnikov Yu.A., Kiryushkin S.G., Mar 'in A.P. *Antioxidation stabilization of polymers* (Russ.) M. Khimiya, 1986, pp. 218-221.

[12] Menson./., Sperling L. *'Polymer blends and their composites'* (Russ.) M. Khimiya, 1979, pp. 241-246.

[13] Reitlinger S.A. *'Permeability of polymeric materials'* (Russ.) M. Khimiya, 1974, 269

[14] Machulis A.N., Tornau E.E. *'Diffusion stabilization of polymers'* (Russ.), Vilmts, MINTIS, 1974, pp. 64-88.

[15] Lipatov Yu.S. *'Physical chemistry of multi-component polymeric system'* (Russ.) Kiev, Naukova Dumka, 1986, v.2, pp. 95-97, 115.

[16] Hiroshi M., Etsuro O. //*Fiber* (Jap.), 1986, v.47, No.2, pp.89-94.

[17] Kiryushkin S.G., *Cand. Diss.*, Moscow, N.N. Nesmeyanov ICP, ANSSSR, 1975.

[18] Popov A.A., Rapoport N.Ya., ZaikovG.E. *'Oxidation of oriented and stressed polymers'* (Russ.) M. Khimiya, 1987.

[19] Bystrov G.A., et al. *'Deactivation and utilization of wastes in plastic materials production' L,* Khimiya', 1982, pp.246-247.

In: Chemical Reactions in Liguid and Solid Phase
Editors: G. E. Zaikov and A. Jiménez, pp. 165-175
ISBN 1-59033-854-5

Chapter 9

SYNTHESIS OF NEW MONOMERS ON DIALLYLGUANIDINE BASIS AND THEIR ABILITY TO RADICAL (CO)POLYMERIZATION

G. E. Zaikov[a], Yu. A. Malkanduev[b], S. Yu. Khashirova[b], A. M. Esmurziev[b], A. I. Martynenko[c], L. I. Sivova[c] and N. A. Sivov[c*].

[a]N.M.Emanuel Institute of Biochemical Physics, Russian Academy of Science, 4 Kosygin st., 117334 Moscow, Russian Federation, Fax: +7 (095) 137 4110, e-mail: chembio@sky.chph.ras.ru

[b]Kh.M.Berbekov Kabardinian Balkarien University, 173 Chernyshevsky st., 360004 Nal'chik, Russian Federation

[c]A.V.Topchiev Institute of Petrochemical Synthesis, Russian Academy of Science, 29 Leninsky prosp., 119991 Moscow, Russian Federation, Fax: +7 (095) 230 2224, e-mail: sivov@ips.ac.ru

ABSTRACT

Methods were elaborated for the synthesis of new monomers, derivatives of diallylguanidine. These monomer salts (acetate and trifluoroacetates) have ability to polymerize in water and organic solvents in presence of radical initiators. Radical copolymerization of diallylguanidine and diallyldimethylammonium chloride was investigated. High biocide ability of obtained copolymers was demonstrated.

Keywords: monomers, radical polymerization, copolymerization, water-soluble polymers, NMR

1. INTRODUCTION

The substances with guanidine groups as it is known exhibit the noticeable biocide activity [1]. The introducing of such groups in polymer chains result in reinforcement of these properties. Many medicines also contain guanidine groups. The purpose of this work was to synthesize diallyl monomers with guanidine groups, to investigate their properties and to ascertain the possibility of the obtaining on their basis of polymer products with biocide activity.

2. EXPERIMENTAL PART

NMR^{1}H spectra were measured using a Bruker MLS-300 instrument in D_2O, $CDCl_3$ at ~ 20^{0}C. IR spectra were recorded with "Spekord M82"(Carlzeis Jena) in KBr tablets. Viscosities were measured with Ubelode-type viscosimeter in 1N NaCl aqueous solution at 30^{0}C.

Solvents were dried by conventional methods and distilled prior to use. DAA was distilled prior to use over NaOH under dry nitrogen, b.p. 109-110^{0}C; NMR^{1}H data are given in Table 1. Cyanamide, acetic acid, trifluoroacetic acid were the commercial products and were used without further purification.

Synthesis of DAGA (1) (typical method was given; the details of syntheses are listed in Table 5). Glacial acetic acid (1.00 mol) was added with stirring to cooled DAA (1.05 mol) in one-neck flask (0.5 L). Then to the resulting mixture of DAA acetate and DAA a solution of cyanamide (1.00 mol) in acetonitrile (~ 60 mL) was added. After this the reaction mixture was stirred for ~10 min without cooling. The prepared mixture was added dropwise to reflux acetonitrile (25 – 250 mL) in a three-neck flask (1.0 L) with a reflux condenser and dropping funnel during ~1h 40 min (crystalline product appeared after addition of half of the mixture), then this mixture was refluxed an additional ~1h 40 min. The precipitate was filtered off, washed many times with acetonitrile, *n*-butanol and acetone, and dried *in vacuo* at ~20^{0}C to give DAGA, m.p. 212 – 214^{0}C. NMR^{1}H data are given in Table 1.

Synthesis of DAG (2). Solution of sodium ethylate was prepared in ethanol (~30 mL of ethanol per 0.1 mol of sodium) and then mixed with ethanol solution of equimolar DAGA (~30 mL of dry ethanol per 0.1 mol of DAGA) with stirring. After addition reaction solution was stirred for 1h 30 min, then kept 12 hours. After filtration off of precipitate of sodium acetate the filtrate was evaporated on a rotary evaporator, yielding DAG, m.p. 166 – 169^{0}C. NMR^{1}H data are given in Table 1.

Synthesis of DAGTFA (3). To obtain this monomer salt the solution of DAG (without base isolation) in ethanol was used. TFAA (2.8 mL, 0.036 mol) was added dropwise to solution of DAG (5.0 g, 0.036 mol) in ethanol (20 mL) under ice cooling, then stirred 3 hour and kept 12 hours. The resulting solution was evaporated on a rotary evaporator, yielding 8.2 g (90%) of DAGTFA, m.p. 156 – 157^{0}C. (Precipitation of this salt from ethanol into dry acetone can be used for it isolation and additional clarification.) When 3-fold excess of TFAA was used monomer salt **3** was obtained in 85% yield, m.p. 156 – 158^{0}C. NMR^{1}H data are given in Table 1.

Table 1. NMR^{1}H Spectral Characteristics of Monomer and Polymer Substances[a,b].

Entry	Substance	CH_3-	-CH_2-N [for polymer: CH_2-N (ring)]	CH_2= [for polymer: CH_2-(chain)]	-CH= [for polymers -CH- (chain)]
			Monomer substances[c]		
1	DAGA	1.91	4.04	5.28, 5.34	5.89
2	DAGTFA	-	4.30	5.57, 5.62	6.16
3	DAG	-	3.88	5.15, 5.19	5.79
4	DAA	-	3.21	5.06, 5.13	5.86
5	DADMAC	3.03	3.91	5.41, 5.45	6.07
			Polymer substances[d]		
6	PDAGA	1.94	3.14, 3.27, 3.54, 3.71	1.12, 1.35, 1.52, 1.73	2.06, 2.39
7	PDAGTFA	-	3.53, 3.84, 3.97	1.26, 1.41, 1.78, 1.97	2.33, 2.67
8	PDADMAC	3.45, 3.55	3.56, 4.12	1.63, 1.85	2.60, 2.99
9	Copolymer of DAGA: DADMAC= 25:75		Signals of DAGA links (lower line)		
		2.17	3.81, 3.97	1.47, 1.70, 1.80, 1.90	2.30, 2.58
		3.43, 3.55	3.58, 4.10	1.62, 1.84	2.64, 3.00
			Signals of DADMAC links (higher line)		

[a] Spectra were measured in D_2O at ~ 20^0C, except entry 4: in $CDCl_3$; chemical shifts δ in ppm. [b] The numbers of protons on integral magnitudes are corresponding to predictable structures. [c] Allyl parts of monomers spectra are $ABMX_2$-type with different multiplicity of their signals; signals of methyl group are singlet ones. [d] All signals are broadened and a number of these signals are partially overlapping with each other.

Table 2. Synthesis of Diallylguanidine Acetate (1)

Entry	DAA,[a] mL (g, mol)	AA,[b] mL (g, mol)	CA,[c] g, (mol)	MeCN,[d] mL	MeCN,[e] mL	t,[f] min	Yield (**1**),[g] %
1	127 (100.5, 1.034)	57 (59.9, 0.986)	41.4 (0.986)	50	25	100 + 110	52
2	169 (133.4, 1.375)	75 (78.6, 1.310)	55.8 (1.310)	65	170	120 + 90	67
3	134 (105.9, 1.070)	59 (62.0, 1.020)	43.9 (1.020)	50	250	100 + 90	75
4	150 (118.7, 1.224)	67 (69.9, 1.186)	49.0 (1.186)	60	250	105 + 105	73
5	157.5 (124.4, 1.280)	70 (73.5, 1.220)	51.3 (1.220)	60	250	105 + 105	75

[a] Quantity of diallylamine. [b] Quantity of glacial acetic acid. [c] Quantity of cyanamide. [d] Volume of acetonitrile for preliminary dissolution of CA. [e] Volume of refluxing acetonitrile for addition of the mixture of DAA acetate, DAA and CA. [f] Reaction time: the first number is the drip time, the second number is the additional refluxing time. [g] Yield was calculated on CA amount

Synthesis of DADMAC. This comonomer was prepared as described before [10]. NMR ^{1}H data are given in Table 1.

NMR ^{1}H data of DAGA and DADMAC homopolymers and their copolymers are given in Table 1.

2. RESULTS AND DISCUSSION

2.1. Syntheses of Monomers

Diallylguanidine (DAG) derivatives were synthesized by reaction of diallylamine (DAA) acetate and cyanamide in acetonitrile (scheme 1) with following transformation of obtained DAG acetate (DAGA, **1**) into DAG (**2**) and DAG trifluoroacetate (DAGTFA, **3**).

Scheme 1

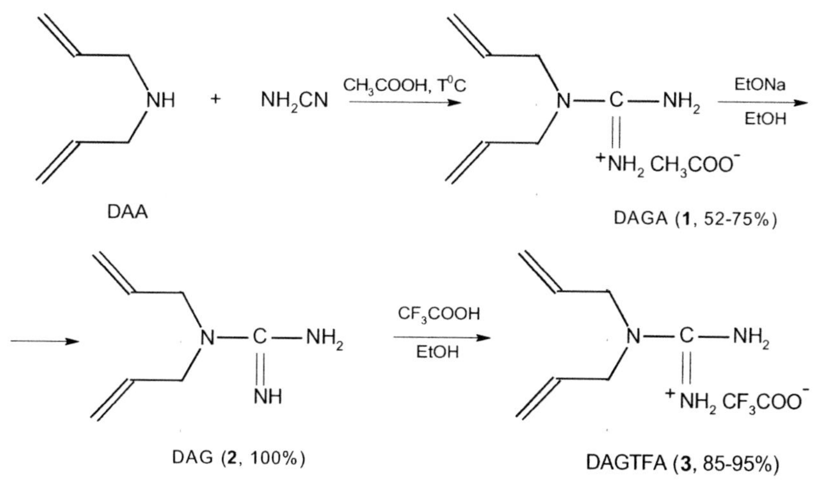

This method is simple and convenient. Preliminary prepared mixture of DAA acetate and DAA was added under cooling to the solution of cyanamide in acetonitrile. Amine is used with some excess because the nucleophilic addition of DAA to cyanamide's nitrile group takes place on the first stage of reaction (scheme 2). The resulting mixture is added dropwise gradually to refluxing acetonitrile. The obtaining DAG (**2**) is unstable at high temperature, but due to greater basicity as compared with DAA DAG come into exchange reaction with DAA acetate and form stable salt – DAGA (**1**), eliminating DAA for following transformation.

Scheme 2

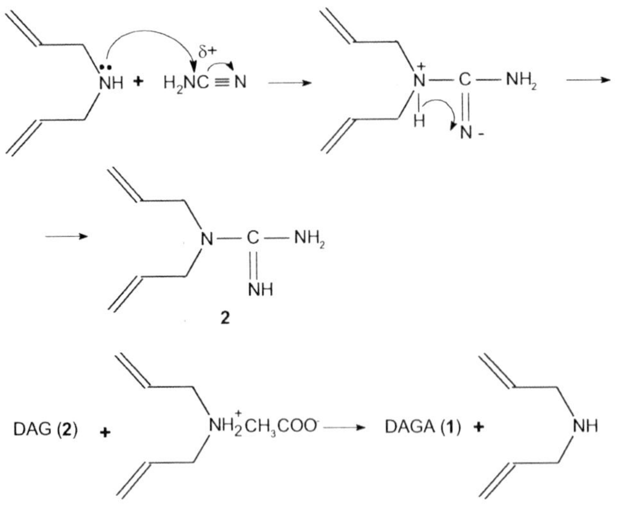

As investigation has shown the volume increase of refluxing acetonitrile (from 25mL to 250 mL per 1 mole of cyanamide) raised the yield of DAGA from 52 to 75%. Probably, protonation reaction of base (**2**) in more dilute solutions occurs with greater rate and yield of salt (**1**) (following dilution will complicate isolation of product).

For above-mentioned reasons DAG (**2**) was obtained quantitatively by salt (**1**) addition into ethanol solution of preliminary prepared sodium ethylate (scheme 1). After sodium acetate separation ethanol solution of DAG (without its isolation) may be used to obtain other salts; such way trifluoroacetate (**3**) was synthesized by the addition of trifluoroacetic acid (TFAA). In wide range of DAG:TFAA ratios (up to 3-fold excess of acid) the salt of equimolar composition was obtained. These were confirmed by data of elemental analysis and melting point measurements. For prepared monomers NMR^{1}H spectra were measured and solubility in water and some organic solvents were determined. Salts solubility changes in the following way:

(1) DAGA – water > methanol > ethanol >> acetone;
(2) DAGTFA – methanol > ethanol > acetone > water.

2.2. Radical Polymerization of DAG Derivative Monomers

Radical polymerization of DAGA (**1**) and DAGTFA (**3**) was studied in water, water-methanol and methanol solutions at different temperatures (20-60^{0}C); ammonium persulfate (APS) and azobisizobutironitrile (AIBN) ([I] = 10^{-2} – 10^{-3} mol L^{-1}) were used as initiator. Polymerization preliminary was shown not to occur without an initiator.

The prepared reaction solutions were degassed in ampoules *in vacuo* (10^{-3} mm Hg) and then sealed and heated. In the case of initiator decay at low temperature (20^{0}C, UV) reaction solution was transferred in quartz cuvette (*in vacuo*).

Polymerization process was controlled by spectra (NMR^{1}H) or viscosity measurements of dialysis cleared and dried *in vacuo* polymeric products.

On the experimental data basis we may suggest that homopolymerization of monomer salts DAGA and DAGFTA are made difficulties under investigated conditions. So, for example, DAGA conversion in polymer in condition ([DAGA] = 4 mol L^{-1}; water; [APS] = $4{\cdot}10^{-3}$ mol L^{-1}; 60^0C; 72 hours) was ~5% ([η] ≈ 0.05 dL g^{-1}. All of these facts indicate on considerable contribution of degradative chain transfer on monomer (CTM) in investigated systems.

It is known [2-4] that reaction of degradative CTM is inherent just for allyl monomers. Due to high α-protons mobility of monomer's allyl groups the abstraction of α-protons by propagation radicals is facilitated to form stable (low active) radicals (scheme 3).

Scheme 3

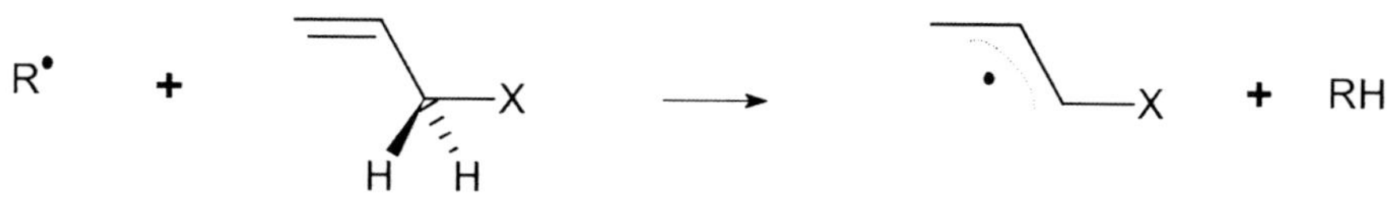

The analogous processes are observed both for dyallylamine and for its N-alkyl derivatives. However monomeric N-alkyl-N,N- diallyl derivatives in quaternized form polymerize easy under radical initiation to obtain cationic high-molecular weight polyelectrolytes [4-6]. Authors [3, 6] have established that both protonation and quaternization of examined monomers result in increase of activation energy of α-protons abstraction due to strengthening α-protons bonds. Because of it considerable inhibition of degradative CTM observe and such CTM replaces by effective CTM [6, 7]. This assumption is confirmed by appearance of the end vinyl and methyl protons signals in NMR^{1}H spectra of corresponding polymers [8, 9] (scheme 4).

Scheme 4

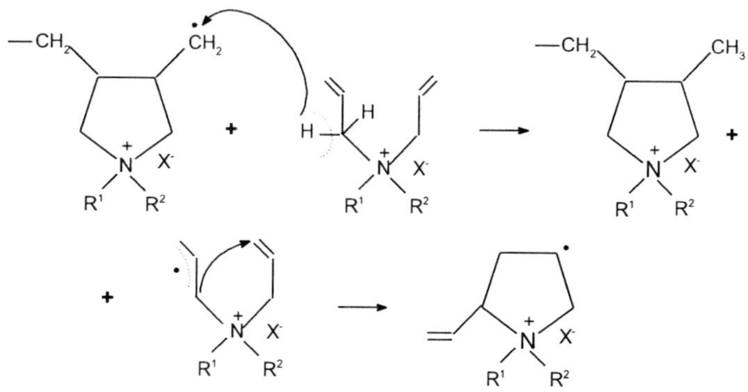

In the investigated systems cationogen monomer salts (**1** or **3**) may exist in aqueous solutions in three resonance structures (I – III) but requirement for suppression of degradative CTM – charge on nitrogen atom bonded to allyl groups – fulfill only in the structure I (scheme 5).

Scheme 5

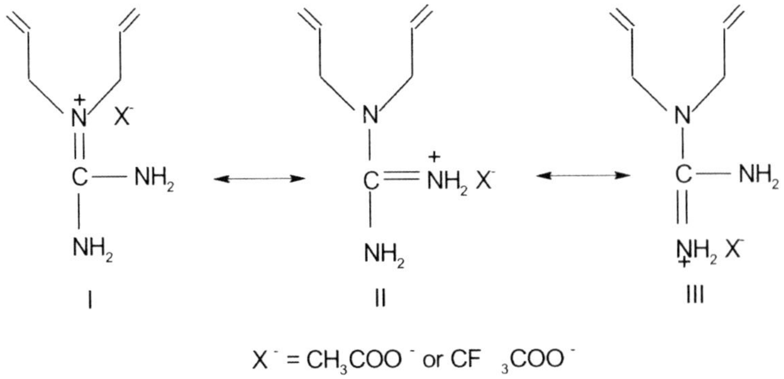

While structures II and III form stable systems (including on steric factors) (scheme 6). Both two nitrogen atoms (IV) and two hydrogen atoms of charged nitrogen (V) generate delocalized structures with acetate (or trifluoroacetate) counterion.

Scheme 6

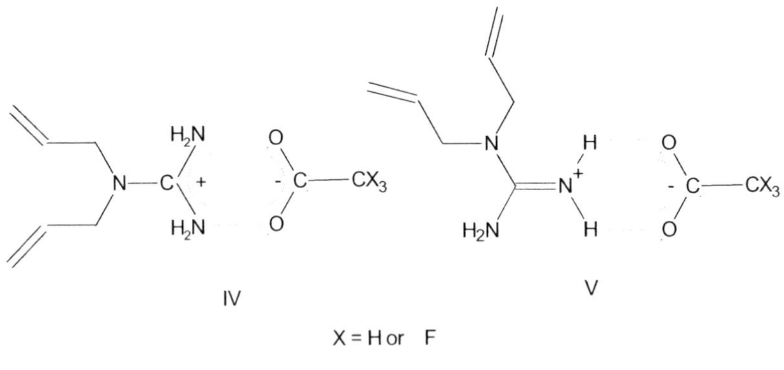

The combined analysis both IR and NMR ^{1}H spectroscopy data confirmed that structure IV is inherent for DAGA and structure V – for DAGTFA. So, $\nu_{C=O}$ (DAGTFA) displays at 1598 cm^{-1} while the same signal for DAGA – at 1579 cm^{-1}; that is in second case delocalization degree in carboxylate ion is greater. The absence of intensive band of C=N bond at 1660 – 1680 cm^{-1} in DAGA IR-spectrum confirms also structure IV for this substance (in DAGTFA IR-spectrum this band is present). The same conclusions can be drew on comparison of CH_2 – N protons chemical shifts of these salts in NMR ^{1}H spectra (see experimental part).

On the basis of these assumptions it can be concluded that for polymerization of monomer salts (**1**) and (**3**) the degradative CTM will be kept in considerable degree. For the case X = F (scheme 6), the decrease of degradative CTM degree would be expected because high inductive effect (-I) of trifluoromethyl group would must cause the decrease of structure V stability. But the expected effects and substantial rate of DAGTFA polymerization couldn't be achieved; probably, because of low monomer solubility in investigated systems (maximum solubility in water 0.5 mol L^{-1} at 60^{0}C).

2.3. Radical Copolymerization of DAGA and DADMAC

Radical copolymerization of DAGA and diallyldimethylammonium chloride (DADMAC) was carried out in ampoules *in vacuo* in aqueous media ([M] = 4.0 – 4.3 mol L^{-1}; initiators [APS], [AIBN] ~ $4{\cdot}10^{-2} - 4{\cdot}10^{-3}$ mol L^{-1}; 60^0C). The comonomers reaction solutions with initiator was degassed (10^{-3} mm Hg) and then sealed and heated. Precipitated copolymers were separated on glassy filter, dried *in vacuo* and twice reprecipitated from methanol in diethyl ether. Copolymer composition was determined on NMR^1H data (spectra were measured in D_2O); intrinsic viscosity was measured on Ubbelhode-type viscosimeter in 1N NaCl aqueous solution at 30^0C.

As a result of accomplished investigations the copolymers broad series of different compositions were obtained in sufficient quality for further physical chemical and biological studying. Copolymerization was carried out up to different conversion degrees (2 – 93%). (Investigation of (co)polymerization to high conversion degrees can bring important results for practical purposes).

In all cases (including low conversion q ≤ 5%) copolymers enriched with DADMAC links in comparison with initial mixture of comonomers were formed (table 3, 4). This fact confirms the greater DADMAC reactivity in propagation reactions. It would be noted that polymerization in solution has taken place in presence of radical initiator only; and it completely has been suppressed by addition of effective radical inhibitors.

Table 3. Dependence of Copolymer Composition on Initial Composition of Reaction Solution for Copolymerization of DADMAC (M_1) and DAGA (M_2) (*[APS] = $4{\cdot}10^{-3}$ mol L^{-1}*; H_2O; 60^0C; Copolymerization 40 hours)

Entry	Initial comonomers mixture		Copolymers		
	$[M]_{sum}$, mol L^{-1}	$M_1 : M_2$, mol%	$M_1 : M_2$, mol%[a]	Q, %[b]	$[\eta]^c$
1	4.00	20:80	30:70	2	0.14
2	4.00	40:60	56:44	3	0.19
3	4.00	50:50	68:32	6	0.21
4	4.26	70:30	93:7	18	0.39
5	4.17	80:20	97:2	31	0.48
6	4.12	90:10	98:2	52	0.64

[a] According to NMR^1H data. [b] Conversion degree. [c] In 1N NaCl aqueous solution at 30^0C, dL g^{-1}.

Thus, on the basis of the obtained data the conclusion was made about comparative reactivity of studying comonomers. These facts were allowed to elaborate methods of synthesis of new cationogen polymer products with given characteristics (composition, molecular weight, hydrophilic-hydrophobic balance), which may be varied in large scale.

It may be more definitely inferred about possible fields of practical usage of synthesized copolymers after additional studying of physical chemical properties and characteristics.

Table 4. Dependence of Copolymer Composition on Initial Composition of Reaction Solution for Copolymerization of DADMAC (M_1) and DAGA (M_2) (*[APS] = $4\cdot10^{-2}$ mol L^{-1}*; H_2O; 60^0C; Copolymerization 33 hours)

Entry	Initial comonomers mixture		Copolymers		
	$[M]_{sum}$, mol L^{-1}	$M_1 : M_2$, mol%	$M_1 : M_2$, mol%[a]	Q, %[b]	$[\eta]$[c]
1	4.00	20:80	38:62	10	0.10
2	4.00	40:60	75:25	23	0.13
3	4.00	50:50	83:17	28	0.17
4	4.29	70:30	87:13	74	0.20
5	4.17	80:20	91:9	80	0.31
6	4.17	90:10	94:6	93	0.52

[a] According to NMR ^1H data. [b] Conversion degree. [c] In 1N NaCl aqueous solution at 30^0C, dL g^{-1}.

Nevertheless, expecting *apriori* considerable bactericide activity of copolymers has been elucidated for a few compositions (table 5). The prepared copolymers demonstrate considerably greater biocide activity than both polyDADMAC and polyhexamethyleneguanidine containing guanidine in each link. It is probably connected with combined action: biocide function of guanidine-containing links and transport function of DADMAC segments of polycation.

Table 5. Biocide Activity of Some Monomers and Copolymers.

Entry	Biocide	MSC, wt.% [a]	
		E-coli	St. aureus
1	DAGA	-[b]	-[b]
2	DAGTFA	$1.5\cdot10^{-2}$	-[b]
3	Copolymer [c]	$1.5\cdot10^{-3}$	$4.0\cdot10^{-3}$
4	Copolymer [d]	$3.5\cdot10^{-4}$	$4.5\cdot10^{-4}$

[a] Minimum suppressing concentration of biocide. [b] No active to corresponding test-bacteria. [c] DADMAC:DAGA = 94:6, $[\eta]$ = 0.52 dL g^{-1} (table 2, en. 6). [d] DADMAC:DAGA = 83:17, $[\eta]$ = 0.17 dL g^{-1} (table 2, en. 3).

Thus, the methods of synthesis of new cationogen monomers DAGA and DAGTFA were elaborated; their physical chemical characteristics were studied; the possibility of involving of these monomer salts in reaction of radical (co)polymerization was shown; and polymer products of different composition and molecular weight were obtained. At the same time the presence of guanidine groups in structure of homo- and copolymers give the opportunity to carry out macromolecular design on their basis both due to polymer analogous and graft modification. Introduced groups with different properties will allow to broaden the field of the practical use of investigating (co)polymers.

REFERENCES

[1] *Chemical Encyclopedia*, ed. By I.L.Knunjanz, Soviet Encyclopedia, Moscow, 1988, 617.

[2] M.Litt, and F.R.Eirich, *J. Polym. Sci.,* 1960, **45**, 379.

[3] V.P.Zubov, M.Vijaya, M.N.Masterova, and V.A.Kabanov, *J. Macromol. Sci. Chem*., 1979, **A13**, 111.

[4] V.A.Kabanov, and D.A.Topchiev, *Vysokomol. soed*, 1988, **30**, 675 [*Polym Sci. USSR*, 1988, **30**, 667 (Engl. Transl.)].

[5] G.B.Butler, *Cyclopolymerization and Cyclocopolymerization*, Marcel Dekker, Inc., New York, 1992, 557pp.

[6] L.M.Timofeeva, Yu.A.Vasil'eva, N.A.Kleshcheva, and D.A.Topchiev, *Izvestia Akademii nauk, ser., khim*., 1999, 865 [*Russ. Chem. Bull*., 1999, **48**, 856 (Engl. Transl.)].

[7] M.F.Hoover, *J. Macromol. Sci. Chem*., 1970, **A4**, 1327.

[8] D.A.Topchiev, G.T.Nazhmetdinova, A.M.Krapivin, V.A.Shrejder, and V.A.Kabanov, *Vysokomol. soed,* 1982, **24**, 473 [*Polym Sci. USSR*, 1982, **24**, 465 (Engl. Transl.)].

[9] Yu.A.Vasil'eva, N.A.Kleshcheva, G.L.Gromova, A.I.Rebrov, M.P.Filatova, E.B.Krut'ko, L.M.Timofeeva, and D.A.Topchiev, *Izvestia Akademii nauk, ser., khim*., 2000, 430 [*Russ. Chem. Bull*., 2000, **49**, 422 (Engl. Transl.)].

[10] J.Nedi, S.Harada, O.Juizuka, *J. Polymer Sci*., A-1, 1967, **5**, 1809.

In: Chemical Reactions in Liguid and Solid Phase
Editors: G. E. Zaikov and A. Jiménez, pp. 177-190
ISBN 1-59033-854-5

Chapter 10

REACTIVITY OF STRESSED MOLECULES: CALCULATION OF THE EFFECT OF CHAIN DEFORMATION ON SCISSION OF MACROMOLECULES AND RADICALS

B. E. Krisyuk
Institute of Problems of Chemical Physics, Russian Academy of Sciences,
Chernogolovka, Moscow Region, 142432 Russia

ABSTRACT

The effect of deformation ε of a polyethylene molecule and radical on the activation energy E_a of the reaction of its monomolecular scission was calculated using the MNDO and AM1 methods. Propane, octane and decane molecules were used as PE chain models. Reactions of molecule and radical scission were shown to be strongly accelerated by deformation, its activation energy, in contrast to the previously studied reactions, being parabolic in chain deformation. The $E_a(\varepsilon)$ dependence for scission of molecules is shown to be linear only in the region of small ($\varepsilon < 3\%$) strains, for degradation of macroradical such region is absent. Both semiempirical methods give close results, but the AM1 method provides a better description of deformation of the macroradical. The proposed approach makes it possible to analytically describe the calculated results, including the case of a reaction proceeding under the influence of a constant force acting on a molecule. The possible cause for difference between the dependences of the local and integral rate constants on the load are considered.

Keywords: Quantum-chemical calculation, deformation, chain scission

INTRODUCTION

Recent experimental results show that the deformation of a reaction center can significantly (sometimes by several orders of magnitude) change its reactivity both in low-molecular-mass compounds [1, 2] and in macromolecules [3-5]. This effect does not depend on the cause of the reaction center deformation, which can be due to an external mechanical field or to structural features of the compound. Presently, the dependence of the rate constant on the external mechanical stress has been studied in detail for the reactions of ozone with C-H bonds of saturated polymers and with double C=C bonds of unsaturated polymers, for the hydrolysis of polyamides and polyesters, for monomolecular scission of strained macromolecules, etc. In spite of the significantly different natures of these reactions, the dependence of the rate constant k on the stress applied σ often assumes the form [3-5]

$$ln\ (k/k_0) = \alpha\sigma/RT \quad (1)$$

where the coefficient α depends on the nature of a reaction, presence of catalysts, and the structure of a material; it determines the deformation susceptibility of the process. Therefore, to predict the response of this or another process to deformation, one should learn to estimate the value of α; for this purpose, one should elucidate the mechanism of the effect.

With this aim in view, we previously performed quantum calculations of the dependence of the activation energy on the deformation of a model molecule for the processes in question. As model molecules, we took low-molecular-mass compounds with the structure of the reaction center similar to that in a polymer. To confine ourselves to a reasonable length of model molecule, we determined the minimum number of its units for which the calculated results become independent of this number. For example, it has been shown that even propane is sufficient to describe the abstraction of a hydrogen atom from PE by ozone [6] and that the PE chain rupture reaction requires consideration of octane at least [7].

The deformation of model molecules was specified by introducing different distances between their terminal skeletal atoms into the problem, without optimization of this parameter in calculations. This approach turned out to be very fruitful. It provided a good description for the reaction of hydrogen atom abstraction from PE and PP by ozone [6, 9] and for acid hydrolysis of polyesters and polyamides [8, 9]. Calculations showed that the activation energy of these processes linearly depends on deformation (or force), in agreement with the experimental dependence [1]. Moreover, a comparison of the calculated results with the experimental data demonstrated a reasonable quantitative agreement between them. However, when we attempted to calculate the tension susceptibility of the monomolecular rupture of the PE chain [7], we encountered certain difficulties: the activation energy of the process decreased faster than in the experiment. This work is an attempt to revert to this problem and approach its solution from a more generalized standpoint. The subject matter of this work is quantum-chemical simulation of the degradation reaction of the macromolecule and middle macroradical. These reactions play an important role in the homogeneous damages in loaded polymers.

METHODS OF CALCULATION

The calculations were conducted using MNDO and, if it was necessary, AM1 Hamiltonian in the MOPAC 7.2 software. The length of the ruptured bond R was used as the reaction coordinate. The deformation was varied by changing the distance (L) between the terminal carbon atoms. In all cases, the chain was in the all-trans planar zigzag conformation.

The model molecules for scission of macromolecule were heptane, octane and decane. Similarly heptyl, octyl, and decyl radicals were used as models of PE middle macroradicals. The ruptured bond was assumed to be in the middle of a molecule (between the third and forth C atoms in the heptyl radical, forth and fifth C atoms in the octyl radical, and fifth and sixth C atoms in the decyl radical), and the free valence was borne by the C atom adjacent to the one involved in the ruptured bond:

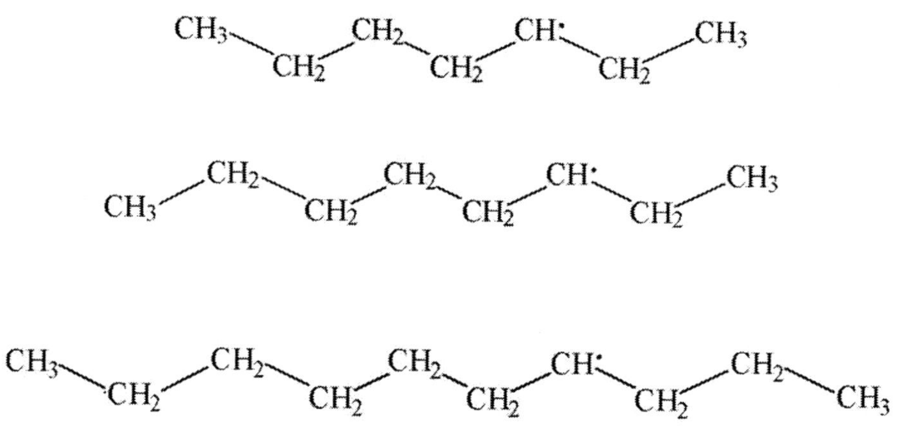

The rupture of this particular bond is the most favorable because this reaction results in the formation of a double bond with a corresponding gain in energy. The calculation proceeded in the following way. First, the equilibrium length of the given model molecule was evaluated. Then, the distance between terminal carbons, corresponding to the specific strain ε from 0 to 5-8%, was fixed. The profile of the potential energy surface was calculated for each ε by variation of the reaction coordinate R with a step of 0.05 A and optimization of all other coordinates.

For the molecular scission the initial state is the ground state of the molecule, the final state is birradical, and it is not known which of the two states is closer to the transition state. Unlike a system composed of two atoms, where the ground state is binding, the birradical state is loosening, and the transient state occurs at the intersection of these two terms, in more complex systems a birradical state can be one of the binding states as well. The well-known stable birradicals provide an example. In this connection, the calculation of the potential energy surface profile was carried out for both ground and birradical states. There is no such problem in reaction of radical scission.

Results and Discussion

1. Deformation

Before passing to the reactions of deformed molecules, let us consider the data on deformation. Fig. 1 shows the dependence of energy of formation for the model molecules and radicals U_0 on specific strain ε or length L. Solid lines depict the calculated points described with parabolas. In all previous calculations [6-9], the MNDO Hamiltonian was used, which provided reasonable results. Whenever possible, the results were compared with the experimental data; good agreement was observed. However, application of this method to calculation of the deformation of the middle radical was not necessarily successful. For example, when describing the deformation of all model radicals, a strict parabolic dependence $U_0(L)$ was obtained at $L > L_0$ (L_0 – length of undeformed molecule), whereas this dependence was hardly obeyed at $L < L_0$. Figure 1a presents the results of this calculation for the decyl radical. Different from linear elastic the dependence $U_0(L)$ at $L < L_0$ does not allow the value of L_0 to be accurately determined. In this connection, in addition to the MNDO Hamiltonian, we used in this work for calculations of radicals reaction the AM1 Hamiltonian, which provided a sufficiently good description of radical deformation (Fig.1).

For both model molecules and radicals Hooke's law well describes of the molecule deformation up to considerable ε, which was previously mentioned in [6-9]. For radicals, despite the above-mentioned disadvantage in parametrization of MNDO method, both methods give close values for stiffness of chain containing a radical fragment (close slopes of curve 1 and 2 in Fig.1a). Thus the dependence of energy U_0 in the vicinity of the initial molecule as a function of length can be described by a parabolic dependence (Hooke's law),

$$U_0(L) = D_0(L - L_0)^2 \qquad (2),$$

where D_0 - the stiffness of the initial molecule (the energy of the initial equilibrium state is assumed to be zero).

Table 1 summarizes the least-squares values of the D_0 and L_0 and calculated from D_0 values of elastic modulus for extended-chain crystal of PE, E (the cross-section $1.8 * 10^{-19}$ m^2 was taken from the X-ray diffraction data [11]). As regards the Young's modulus E for molecule thus obtained, it is almost the same for both octane (290 Gpa) and decane (296 Gpa). The Young's modulus for a heptane molecule is 257 GPa, which implies that this molecule is small and cannot serve as a model of the polymer chain deformation. The calculated values of E are close to the known experimental data and the results obtained by other calculation methods that are typically around 270 GPa. The cause of uncertainty in the E value for the polymer chain and some scatter of the values, obtained by different methods, is discussed in [12]. We will only note here that the use of octane as a model molecule provides an adequate description of the deformation of the PE macrochains.

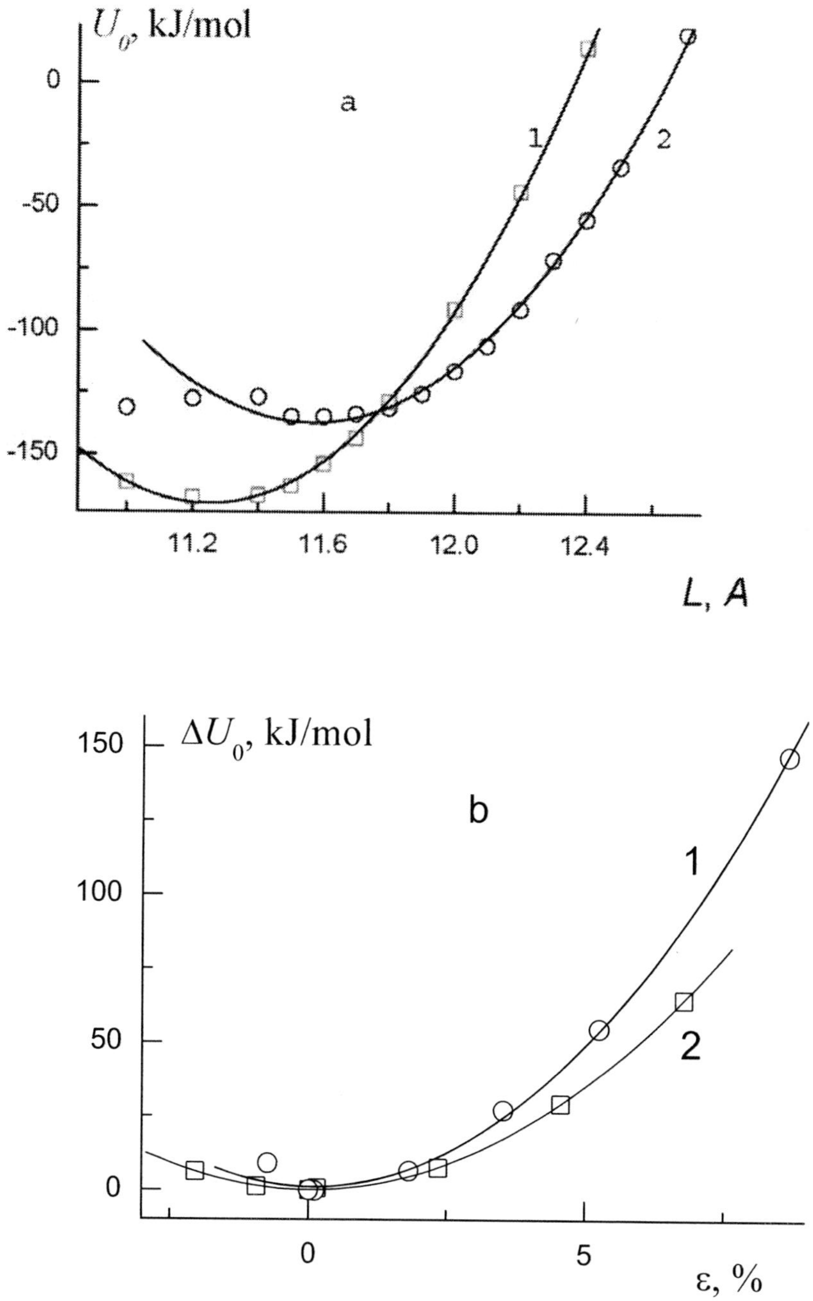

Figure 1. The dependence of energy of formation U_0 for decyl (a) on radical length L (1 – AM1, 2 – MNDO data) and changes in U_0 (b) for octane (1) and decane (2) molecule vs. strain ε (calculated using MNDO Hamiltonian).

Table 1. The Values of Equilibrium Length L_0 and Energy of Formation U_0^0 of Undeformed Model Molecules and Radicals, their Stiffness D_0 and Elastic Modulus E, Obtained Using MNDO and AM1 (In Brackets) Hamiltonians.

Model molecule	L_0, A	$-U_0^0$, kJ/mol	D_0, kJ/(mol*A^2)	E, GPa
Heptane	7.77	183.5	177.1	257
Octane	9.09	202.1	175.7	290
Decane	11.67	241.6	137.7	296
Heptyl	7.66 (7.47)	77.3 (84.3)	156.1 (176.5)	221 (244)
Octyl	9.01 (8.75	95.5 (113.3)	156.1 (178.8)	260 (282)
Decyl	11.57 (11.30)	137.1 (169.3)	120.9 (138.4)	258 (287)

For model radicals the values of E obtained using the MNDO method are 10% smaller than those obtained with the AM1 Hamiltonian. This can be considered as a good agreement. For the heptyl radical, as well as for heptane molecule, the obtained elastic modulus is smaller than for other radicals. Therefore, this model molecule is too small for the description of deformation and the results obtained for the heptyl radical can contain artifacts. The values of E for octyl and decyl radicals coincide with each other, being close to the elastic modulus of a PE chain equal to 250-270 GPa (both experimental and calculated values [12]). The MNDO-calculated elastic modulus of a PE chain is equal to 290 GPa (Table 1). Comparing the MNDO values obtained for the radical and the molecule, we can conclude that the presence of a radical fragment leads to a decrease in the elastic modulus of the chain by about 10% (Table 1). Hence, calculations using MNDO and AM1 semi-empirical methods give us reasonable results on deformation of both macromolecules and middle macroradicals.

2. Scission Reaction

It should be emphasized that a connection between activation energy and deformation is quite natural for the given reactions: the more energy "pumped" in the bond elongation, the less energy has to be added for its rupture. However, it is not the bond, but the whole molecule, that is being strained; this is manifested, for instance, by a deformation relaxation of the free radicals evolved. That is why, for more accurate description of the potential energy surface (PES) profile in the region of the transition state, the calculation was carried out with a fixed R and increasing L, whereby a true reaction coordinate is represented by a combination of the two parameters. It should be also noted that we always refer to PES in R - L coordinates without subtracting the energy of the initial state at a given strain. The reason for this consists in that the most important parameters for a description of reactions of the deformed molecules are the coordinates of the saddle point on this surface (R^*, L^*), i.e., the point through which the reaction of a free molecule proceeds, the parameter L^*-L_0 is a sought sensitivity to strain for reaction under investigation [7, 8].

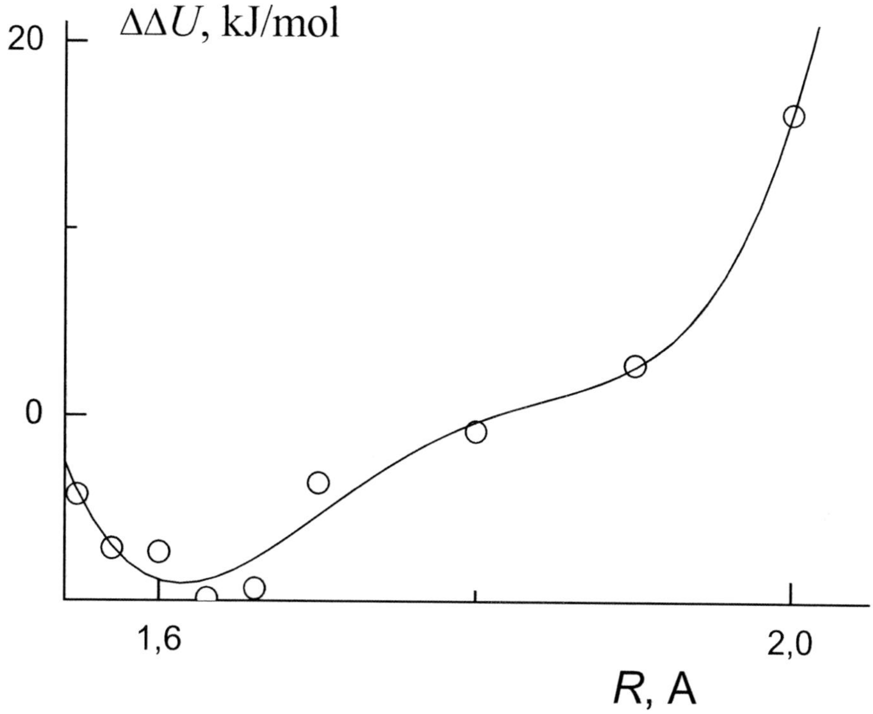

Figure 2. The difference of the formation energies of the ground and biradical states *ΔΔU* as a function of ruptured bond length *R* for octane molecule.

Fig. 2 shows the difference between the calculated values of standard energy of formation of the ground and biradical states *ΔΔU* for the octane molecule, plotted against the *R* value. As expected for the C-C bond length close to the equilibrium value (1.548 Å according to the MNDO data), the ground state is more favorable in energy than the biradical, as confirmed by negative values of *ΔΔU* (the latter state is binding as well). On reaching R = 1.87 Å, the values of the energy of formation of both states become equal, and the biradical state becomes more preferable at large *R*. It is important to note that intersection of the energy levels is fairly smooth, as indicated by a rather small slope of the curve at the abscissa axis intersection (Fig. 2) as compared to the slope of the *ΔU(R)* dependence at the same *R* value (35.6 vs. 503 kJ/(mol*A), respectively). This testifies to the fact that the terms are passing close to each other for a rather wide range of *R*. Therefore, one can assume that nothing hinders this transition in a real reaction.

Hence, the PES profile of the reaction macromolecule scission must incorporate the transition from the ground state to biradical prior to the formation of the transition state and the latter is biradical.

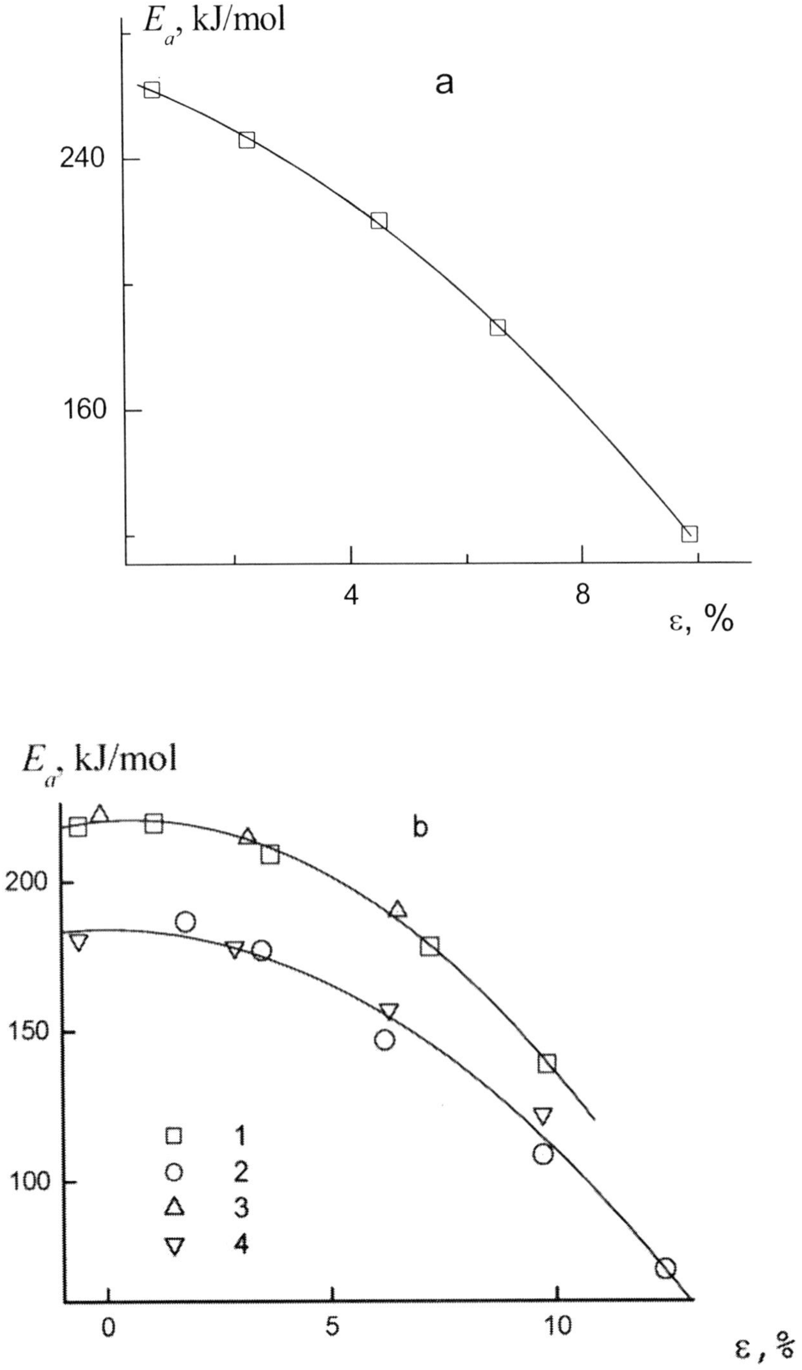

Figure 3. The activation energy of scission E_a for octane (MNDO results, a) and model radicals (b) versus strain ε. Results of calculations using MNDO (1, 2) and AM1 (3, 4) methods for decyl (1, 4) and octyl (2, 3) radicals.

Table 2. The Energies of Formation ΔU for the States at the Crest of the Ridge of the Potential Energy Surface at Different R and L Values Near the Saddle Point (The Model Molecule is Octane)

R, A	*L, A*	*ΔU, kJ/mol*	*R, A*	L, A	*ΔU*, kJ/mol
	L = const			*R* = const	
2.4	9.45	43.53	2.65	9.15	59.25
2.5	9.53	47.72	2.68	9.30	53.80
2.6	9.63	49.02	2.67	9.50	50.15
2.7	9.72	47.14	2.59	9.70	49.42
2.8	9.92	43.45	2.37	10.00	60.38

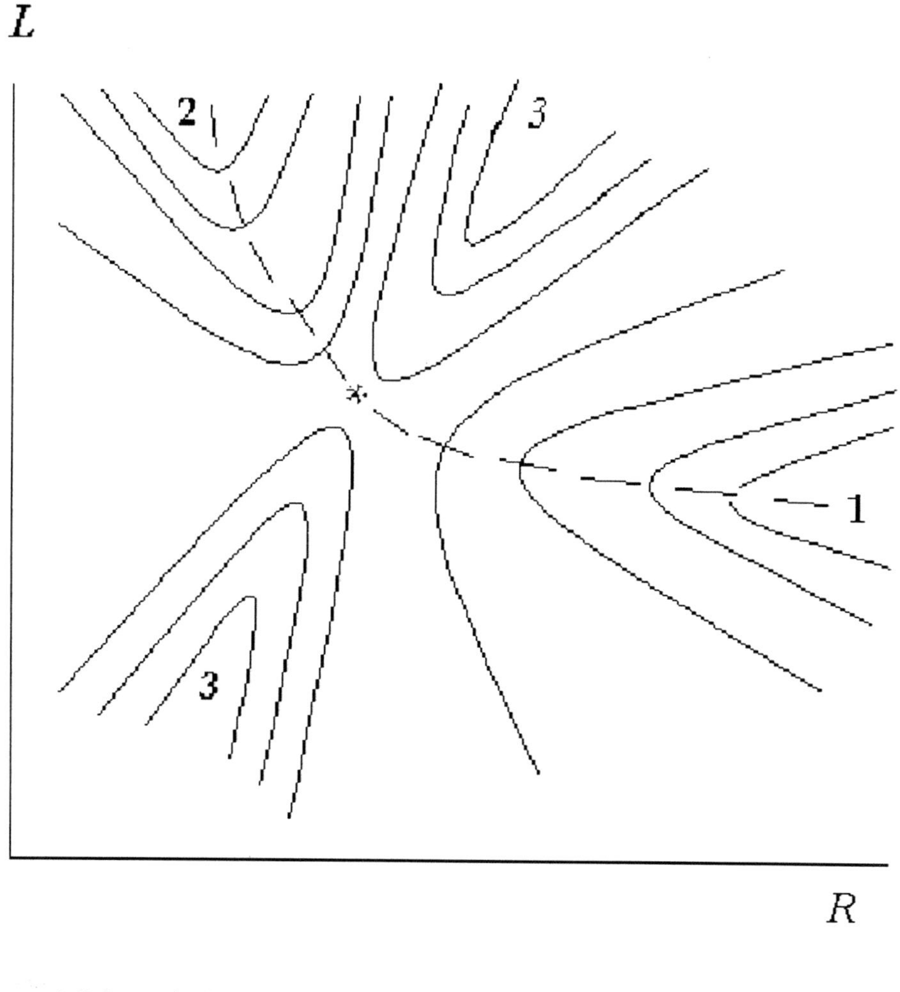

Figure 4. Schematic sketch of the potential energy surface for the reaction studied as a function of R and L: 1 – reactant valley, 2 – product valley, 3 – ridge. Asterisk shows the saddle point, dashed line indicates the reaction path for a free molecule.

Fig. 3 shows the dependence of the change in the activation energy ΔE_a of the rupture of a PE molecule and middle radical on its deformation. This dependence is nonlinear. Fig. 3 demonstrates that it is adequately described by a parabola, but this does not agree with numerous experimental data. In this work, the calculation is performed as the cross section of the potential energy surface at a fixed length. But, in this case, the length L itself becomes a rightful reaction coordinate, as is confirmed by calculations at a constant R and increasing L. Another cross section of the same potential energy surface is thus obtained. The results of such calculations for the reaction of macromolecule degradation exemplified by octane as a model molecule are shown in Table 2. Apparently, both calculation variants yield close results and show that the potential energy surface of this reaction has a saddle point, just as in the case of other processes [6-9]. The shape of the potential energy surface for the process in question is schematically presented in Fig. 4. The approach proposed in [6] provides analytical description for the calculated results. Since the energy as a function of two coordinates $U(R, L)$ has a saddle point, the potential energy surface at small deviations of these coordinates from the saddle point can be described by the quadratic form:

$$U^*(R, L) = A(L - L^*)^2 + B(L - L^*)(R - R^*) + C(R - R^*)^2 + E_a^0 \quad (3)$$

Here, R^* and L^* are the coordinates of the saddle point, E_a^0 is the activation energy of the reaction for the free molecule, and A, B, and C are the interpolation coefficients obtained by least-squares processing of the calculated results. If the reaction takes place at $L = L' =$ const, it will pass a point with the coordinates (L', R') rather than the saddle point and the R' value will be determined from the relationship

$$\partial U^*(R, L')/\partial R = 0,$$

whence $R = R^* - (L - L^*)B/2C$. Substituting this expression into formula (3), we obtain

$$U^*(L') = D^*(L' - L^*)^2 + E_a^0.$$

where $D^* = (A - B^2/4C)$ has the physical meaning of the stiffness of the transition state. Then, allowing for the fact that $f = 2D(L'-L_0)$ is a force acting on the molecule prior to the beginning of the reaction, the change in the activation energy for the relative deformation of the reaction center $\varepsilon = (L - L_0)/L_0 = const$ can be described by the expression

$$\Delta E_a = -f^2(D_0 - D^*)/4D_0^2 - f(L^* - L_0)D^*/D_0 + D^*(L^* - L_0)^2 \quad (4)$$

The free term in Eq. (4) corresponds to an increase in the activation energy due to the fact that the reaction passes a barrier at $L = L_0$ rather than at L^*. A quadratic dependence of ΔE_a on f and, accordingly, on ε appears in this expression. Earlier [6-9], we confined ourselves to the linear term in such an analysis. As was mentioned above, the calculation was performed at L = const or at R - const. However, the real conditions of macrochain loading in a polymer are more adequately represented by the regime $f =$ const, where f is a force acting on the reaction fragment of a macromolecule. This regime corresponds to the reaction path way over the potential energy surface with a constant energy gradient. In this case, the length of the reaction fragment will change during the reaction from $L' = f/2D_0 + L_0$ in the initial state to $L'' = f/2D^* + L^*$ in the transition state. Then, similarly to the above calculations, one can easily derive an expression describing the change in the activation energy at $f =$ const:

$$\Delta E_a = -f^2(1/D^* - 1/D_0)/4 - f(L^* - L_0) \quad (5)$$

The values of the parameters of Eqs. (4) and (5) are presented in Table 3. Let us consider some principal properties of these equations.

The formation of a transition state is usually accompanied by a certain loosening of atom-atom bonds. For the reaction of backbone bonds of macromolecule this means that $L^* > L_0$ (which is usually the case). Then $\Delta E_a < 0$ for any deformations; that is, tension activates the process. If $L^* < L_0$, $\Delta E_a > 0$ at small deformations and $\Delta E_a < 0$ at large ones. Thus, the pattern of the $\Delta E_a(\varepsilon)$ dependence can change with increasing ε for such processes: from inhibition of the reaction to its acceleration.

If the stiffness increases during the formation of the transition state, the $E_a(\varepsilon)$ dependence at $L^* < L_0$ will be a monotonically increasing function for all ε; $L^* > L_0$, $E_a(\varepsilon)$ will decrease at small ε values and increase at large ones. If the formation of the transition state does not entail any significant loosening of skeletal bonds, that is, $D^* \approx D_0$, then expressions (4) and (5) coincide with an accuracy up to the free term and the dependence of ΔE_a on the force or deformation becomes linear. The force susceptibility does not depend on the reaction pathway (L = const or f = const); it is the case considered earlier [6].

If $D^* \neq D_0$, dependences (4) and (5) are also linear in the region of small deformations, but the force susceptibility changes as the reaction path changes. The region of small deformations itself is limited by the ε_c value, which, as follows from both formulas, is determined by the relationship

$$\varepsilon_c << 2D^*(L^* - L_0)/\{L_0(D_0 - D^*)\} \quad (6)$$

If the length of the reaction center does not significantly change during the formation of the transition state, that is, $(L^* - L_0) = 0$, then the linear part will be completely absent ($\varepsilon_c = 0$) and the dependence of the activation energy on the deformation or force will be parabolic. If the condition $D^* \approx D_0$ is also satisfied, the reactivity of such systems does not depend on deformation.

The studied PE chain breakage reaction is a good illustration of these conclusions. For this reaction the value ε_c is fairly high and amounts to 5.4%, because the parameter $(L^* - L_0)$ = 0.6 A is very high. The parameters of the initial and transition states for this reaction are presented in Table 1, 3. At small deformations, the $\Delta E_a(\varepsilon)$ dependence is virtually linear (Fig. 3), but subsequently the quadratic term also manifests itself.

For the reaction of middle radical scission (Fig. 3) both MNDO and AM1 methods give a parabolic dependence of $\Delta E_a(\varepsilon)$ from the very beginning. The points obtained by the same method for octyl and decyl radicals are fitted by a single curve. This indicates that even the octyl radical perfectly reflects the specifics of the reaction of longer molecules and, thus, appears to be a representative model for the macroradical degradation reaction.

Tables 1 and 3 show the values of L_0 and L^* almost coincide for both octyl and decyl radicals. According to Eq. (5), this must result in the absence of the dependence of E_a on ε if there is a weak change in the molecular stiffness D upon the formation of the transition state. But in our case such assumption is not valid: the transition state is characterized by 4 to 5 or 3 to 3.5 times smaller chain stiffness; according to the MNDO and AM1 method, respectively.

If L_0 and L^* are close, the linear term in Eq. (5) is zero and dependence $\Delta E_a(\varepsilon)$ is determined by the quadratic term. It is this case that is considered in the present work. Therefore, for the reaction of middle radical scission, the region of small deformations in which the quadratic term can be ignored (as was done in the case of other reactions), does not exist.

Table 3. The Values of the Ruptured Bond Lenth R^*, the Model Molecule Length L^*, the Stiffness D^* and the Energy of Formation U_0^* for the Transition State and the Values of Activation Energy E_a^0 for Free Molecules and Radicals Scission. Numerator – MNDO Data, Denominator – AM1.

Model	R^*, A	L^*, A	D^*, kJ/(mol*A^2)	U_0^*, kJ/mol	E_a^0, kJ/mol
Octane	2.60	9.63	57.82	48.9	232.4
Decane	2.57	12.27	47.51	18.4	220.5
Heptyl	2.17/2.13	7.69/7.59	55.6/111.0	142.3/60.2	219.6/144.5
Octyl	2.17/2.13	8.93/8.70	52.5/90.4	124.4/68.4	219.9/181.7
Decyl	2.17/2.13	11.61/11.44	61.9/97.5	85.1/25.1	222.2/194.4

Thus, the linear dependence of the local rate constant on the load is observed only in some cases. There can be reactions that pass from the linear to the quadratic dependence as the deformation increases; in some cases, this dependence is quadratic from the very beginning. But the dependence of type (1), that is, a linear $\Delta E_a(\varepsilon)$ dependence, is observed in all studies of this phenomenon. The absence of a quadratic dependence in experiments can be due to several reasons.

First, let us estimate the real deformations that a macromolecule can undergo in a loaded polymer. The chain elasticity modulus in PE amounts to ≈ 250 GPa, depending on the length of the model molecule. A highly oriented high-density PE can bear a load up to 1000 MPa. If this load is uniformly distributed over all the chains, their deformation will be only 0.36%. For a tenfold overload (that is, for the situation where only 10% of the chains bear the whole external load), the deformation of the load-bearing chains increases to 3.6%. This result should be regarded as an upper estimate, because the values used in calculations were intentionally chosen to be equal to the limiting values for ordinary (not ultra-high-strength) polymers. Actually, real loads are significantly (one or two orders of magnitude) smaller. Then, the deformation of macrochains seldom exceeds 1%; that is, in most cases, reactions proceed at $\varepsilon \ll \varepsilon_c$ and, therefore, fall into the region of a linear $\Delta E_a(\varepsilon)$ dependence.

In the general case, the macroscopic rate constant is a convolution of the local constant $k(f)$, where f is the local force, with the distribution function $F(\sigma, f)$ of the external stress σ over the polymer chains:

$$k(\sigma) = \int F(\sigma, f)\, k(f) df \tag{7}$$

The pattern of the distribution function $F(\sigma, f)$ is unknown, but there are some approaches to its estimation. It is assumed that the majority of chains do not bear any load but the whole external load is distributed among 10% of the bonds. Accordingly, the $F(\sigma, f)$ function must have the form of a steeply decreasing function f with a long "shoulder" at high stress. The contribution of each group of differently loaded chains to the integral rate constant depends on the load susceptibility of the local constant of the reaction in question. For example, in the case of reactions whose rate constant is highly sensitive to the chain deformation, the contribution of the most strained bonds is maximum; in the case of reactions with a low susceptibility to deformation, the reaction of unloaded chains can make a significant contribution to the integral rate constant. Thus, the main contribution to the integral rate constant for each specific reaction is due to its relevant groups of strained chains, that is, to different portions of the distribution function. Hence, the local rate constant for different reactions is actually integrated in formula (7) over different distribution functions. As a result, the integrated dependence $k(\sigma)$ can have the same pattern even for different dependences of the local rate constants on the force. In addition, if the reaction is accompanied by the rupture of macromolecules, the load is redistributed among the intact chains during the process and, accordingly, the pattern of the $F(\sigma, f)$ function changes. This complicates the problem even further. The only evident thing is that the $k(f) \sim \exp(f^2)$ dependence at a definite pattern of the $F(\sigma, f)$ function can also yield the equation of form (1) for the integral rate constant.

Finally, the third reason for the experimentally observed linear relationship between $\log(k)$ and σ can be that the researchers took the deviations from linearity for inflexion of this dependence that are due to changes in the chemical mechanism of the process (see, for instance, rev. [13]). Indeed, it is virtually impossible to distinguish two linear segments and a parabola if the experimental range of σ values is sufficiently narrow.

CONCLUSIONS

Our analysis showed that there are reactions for which the activation energy of the local rate constant depends nonlinearly on the force (for example, the reaction of the monomolecular decay of the PE chain considered in this work). There exists only narrow region of small deformations, where the deviation from linearity is negligible, is shown to exist for these reactions; this occurs often but not always. According to formula (6), this region does not exist (that is, $\varepsilon_c = 0$) for reactions where $(L^* - L_0) \cong 0$. In this case, the linear portion is totally absent and the dependence of the activation energy on the force for such reactions is quadratic up from the smallest deformations.

In this work, the similar dependence of the activation energy for the degradation of the middle radical on the force acting on the chain (deformation) has been obtained. This relationship is quadratic, with the coefficient at linear term being practically zero. This results in absence of the region of small deformations in which the dependence can be approximated by a linear function. This type of dependence is explained by a considerable variation of the stiffness and a small change in the chain length upon the formation of the transition state.

The search for such processes and their study will be the aim of further research. We expect that their revelation as a result of calculations and an analysis of experimental data will enable us to make some conclusions as to the pattern of the $F(\sigma, f)$ function, because this function must satisfy a requirement that is contradictory at first sight: its convolution with $\exp(f)$ for one reaction and with $\exp(f^2)$ for another reaction should yield $\exp(\sigma)$ in both cases, as follows from the experimental data available at present.

REFERENCES

[1] Johes, S.H. and Whittle, E., *Int. J. Chem. Kinet.*, 2:6 479-481 (1981).
[2] Popov, A.A. and Zaikov, G.E., *J. Macromol Sci., Rev. Macromol Chem.*, 27:3-4 343-378, 379-392 (1988).
[3] Krisyuk, B.E., Popov, A.A., and Denisov, E.T., *Vysokomol. Soedin., Ser. A*, 30:8 1736-1740, 1741-1745 (1988).
[4] Krisyuk, B.E. and Cheremisin, V.V., *Polymer Science*, 34:11 975-977 (1992)
[5] Krisyuk, B.E. and Smirnov, K.L., *Vysokomol. Soedin., Ser. A*, 31:2 328-332 (1989).
[6] Krisyuk, B.E. and Polianchik, E.V., *Khim. Fiz.*, 9:1 127-134 (1990).
[7] Krisyuk, B.E., *Polymer Science, Ser. A*, 39:3 340-344 (1997).
[8] Krisyuk, B.E. and Polianchik, E.V., *Khim. Fiz.*, 12:2 253-259 (1993).
[9] Krisyuk, B.E. and Polianczyk, E.V., *Int. J. Polym. Mat.*, 23:1-2 1-16 (1993).
[10] Timerghazin, Q. K., Khursan, S. L. and Shereshovetc, V. V., *J. Molec. Struct. (Theochem)*, 499:1 87-93 (1999).
[11] Sakurada, L, Ito, T., and Nakamae, K., *J. Polym. Sci., Part C: Polym. Lett.*, 15:1 75-80 (1966).
[12] Sukhai, S., *J. Chem. Phys.*, 84:9 5071-5076 (1986).
[13] Manelis, G.B. and Polianchik, E.V., *Sov. Sci. Rev. B. Chem.*, 15:1 61-101 (1991).

In: Chemical Reactions in Liguid and Solid Phase
Editors: G. E. Zaikov and A. Jiménez, pp. 191-196
ISBN 1-59033-854-5

Chapter 11

ARABINOGALACTAN. OXIDATIVE DESTRUCTION*

***Yu. B. Monakov*[a,b], *I. M. Borisov*[b], *G. A. Tolstikov*[a], *E. N. Shirokova*[a] and *G. E. Zaikov*[c]**

[a]Institute of Organic Chemistry USC of Russian Academy of Sciences, 71 Pr. Octyabrya, 450054 Ufa, Russia
[b]Bashkir State University, 32 Ul. Frunze, 450074 Ufa, Russia
[c]Institute of Biochemical Physics Russian Academy of Sciences 4 Kosygin Street, Moscow, 117334, Russia

ABSTRACT

The regularities of the oxidative destruction of the natural polysaccharide-arabinogalactan- in the aqueous solutions under action of hydrogen peroxide and the molecular oxygen have been investigated. It was shown that the state of the oxidative destruction of the biopolymer increases with the growth of the initial concentration of hydrogen peroxide in the solution.

INTRODUCTION

Arabinogalactan (AG) of the larch wood (poly[β-L-Ara.p-(l-6)-β-D-Gal.p-(l-6), -β-D-Gal.p-(l-6)--β-D-Gal.p-(1-6),-β-L-Ara.f-β-D-Gal.p-(1 -6),-β-D-Gal.p-(1 -6)--β-D-Gal.p-(1 - 6)], is the water soluble biologically active representative of the natural polysaccharides [1-3]. It is considered at present time as the potential polymeric matrix for farm products of prolonged action, and also as the additive in the food and cosmetic industries. It is known [4-6] that the important property of the natural polysaccharides (dextran, chondroitin sulphate, chitosan) at their practical application is their molecular mass. Because of this the

* The work has been conducted under a partial financial support of the Project VP 05.01.024 on the scientific program 'Fundamental Investigations of the Higher Education Schools in the area of natural and humanitarian sciences: Russian Universities'

polysaccharides undergo destruction and than, they are applied as the water soluble oligomers. To obtain the degraded polysaccharides, the alkali, acidic and the fermentative hydrolysis [2] and the oxidative destruction under action of ozone [7] and hydrogen peroxide [8, 9] are used. In the case of acidic or alkali hydrolysis of biopolymers, the problem of the obtained products purification from the acid or alkali residues occurs. This will significantly complicate the process and increase the product mill cost. The ozonolithic and peroxide methods are free from the discrepancies and allow to obtain the ecologically clean oligomers. At the present time, the mechanism of the liquid-phase oxidation and destruction of arabinogalactan to obtain the biologically active products with a regulated molecular mass has not practically investigated. Because of this, in this work, the regularities of the oxidative destruction of arabinogalactan in aqueous solutions under action of the system 'hydrogen peroxide-molecular oxygen' has been studied. Two tasks are being solved. Firstly, the given oxidative system was used to regulate the state of destruction of AG. Secondly, in the course of the oxidative destruction, the AG oligomer functional ization occurs, what widens their ability to be used as the complex former of the medical drugs.

EXPERIMENTAL PART

As the object for investigation, the arabinogalactan with molecular mass 40,000 has been used. It was prepared by the water extraction from the larch needles.

The experiments were conducted in the reactor of the gas passing through type. The samples were periodically taken from the reactor. After the reaction finish, the polymeric fraction of the oxidized AG was separated from the oligomeric one by acetone precipitation. The oligomer fraction was separated from the liquid above the precipitate by removing acetone and water.

The analysis of the functional groups of the polymeric and oligomer fractions was conducted by using the method of IR-spectrometry (Vaseline oil, spectrophotometer UR-20). The concentration of the peroxide reaction products was determined by iodometry.

The kinematic viscosity of the 10% aqueous solutions of AG was determined in the capillary viscometers with the hung level at temperature 90±0.5 °C. The characteristic viscosity was determined by using the aqueous solutions of AG + 0.30 M NaCl at temperature 30 °C. The values of the characteristic viscosity [η] in dl/g were calculated by the method of the double extrapolation of the dependencies η_{intr}/C and ln η_{rel}/C to the zero concentration, where η_{intr} and η_{rel} are the intrinsic and the relative viscosities correspondingly, and C is the solution concentration in g/dl.

RESULTS AND DISCUSSION

The thermal destruction of polysaccharide does not observed at heating the aqueous solutions of arabinogalactan in the temperature range 50 °C to 95 °C. In this temperature range, arabinogalactan does not have destruction also at the air passing through the AG aqueous solutions. At hydroperoxide addition to the AG aqueous solutions containing dissolved O_2, the kinematic and characteristic viscosities are reduced (Fig.l and Tables 1 and 2).

The state of the viscosities reduction (Tables 1 and 2) increases with the initial concentration of H_2O_2 in the aqueous AG solutions increase. The initial concentration of the hydrogen peroxide in the system $AG+H_2O+H_2O_2+O_2$ affects also the yield of the oligomer and polymer fractions of the polysaccharide (Table 3).

However, the consideration of the dynamics of the kinematic viscosity change in the course of the reaction (Fig.l) indicates on the fact that the destructive processes especially visible at low concentrations of hydrogen peroxide ($[H_2O_2]_0 < 1$ mol/1) in the AG solution. The small reduction of the kinematic viscosity at high enough values of the initial hydrogen peroxide concentrations in the solution (Fig. 1) can be explained as follows. It is possible that at heating of the reaction mixture $AG+H_2O+H_2O_2+O_2$, the intermediate compounds are formed. The compounds are able to react with each other (crosslinking) what leads to the retardation of the kinematic viscosity reduction. In the crosslinked macromolecules, the functional groups same as in the original AG are retained. Because of this, the groups are also involved in the destruction reactions at the deep stages (Fig.l).

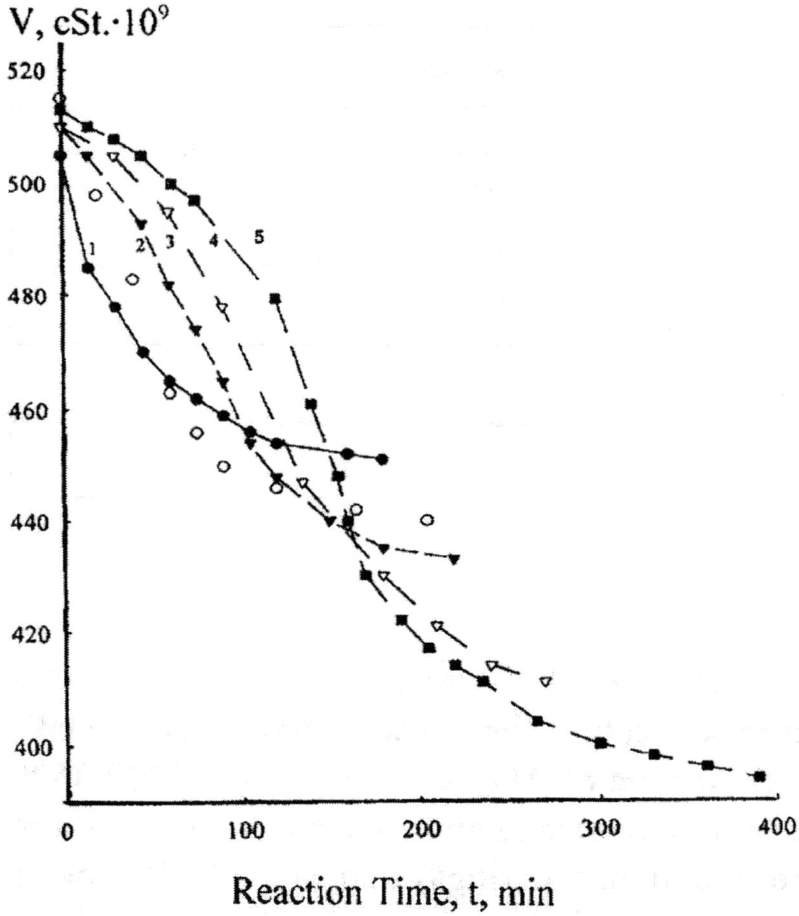

Figure 1. Dependence of the kinematic viscosity, *V*, on the reaction time, *t* at different concentration of hydrogen peroxide in the arabinogalactan [AG] aqueous solutions: 1) 0.5, 2) 0.8, 3) 1.0, 4) 2.0, 5) 2.5 mol/1. T=90 °C $[AG]_0$ =10 mass %

Table 1. Reduction of the Kinematic Viscosity, V, of the Arabinogalactan Solution at Different Initial Hydroperoxide Concentrations. $[AG]_0$=10 Mass %, 90 °C, 6 Hrs, V_0 and V_{ro}-the Kinematic Viscosities of the AG Solution in the Beginning and at the End of the Test, Correspondingly.

Relative kinematic viscosity	Hydroperoxide concentration, $[H_2O_2]_0$, mol./l				
	0.5	0.8	1.0	2.0	2.5
(Vo-VJ/Vo x 100,%	11	15	16	20	24

It should be especially emphasized that the AG destruction is observed only in the case, when the hydrogen peroxide is consumed in the presence of the dissolved oxygen in the polysaccharide solution. The biopolymer destruction is accompanied by accumulation of the carbonyl groups on the macromolecules and the forming oligomers (Fig.2)

Table 2. Reduction of the Characteristic, N, of the Arabinogalactan Solution at Different Initial Hydroperoxide Concentrations. $[AG]_0$=10 Mass %, 90 °C, 6 Hrs, $[N]_0$ =0.033 Dl/G, $[N]_{OT}$-the Characteristic Viscosities of the AG Solution at the End of the Test.

Relative characteristic viscosity	Hydroperoxide concentration, $[H_2O_2]_0$, mol./l			
	0.10	0.20	0.65	1.10
([nL>-[nU/[n]o)xioo,%	0	3	9	15

Table 3. Yield of the Oligomer and Polymer Fractions of the Destructed Arabinogalactan at Different Hydroperoxide Concentrations. $[AG]_0$ = 10 Mass %, 90°C

[HAL mol/l	Yield of fractions, mass %	
	Oligomer	Polymer
0.8	54	42
1.0	73	24
2.0	85	11
2.5	89	4

These facts indicate that simultaneously with the rnacromolecules destruction, the oxidative processes also take place. At the IR-spectra of AG, the vibrations at 3200-3600 cm^{-1} (HO-groups) as the strong and wide band appear. There are also absorbency bands at 1050-1100 cm^{-1} (pyranose and furanose rings) and at 890-910 cm^{-1} (deformation vibrations of C_1-H). The existence in arabinogalactan of the uronic acid residues leads to appearing of the absorbency band at 1690 cm^{-1} of a modest intensity, and of the small intensity band at 1730 cm^{-1}.

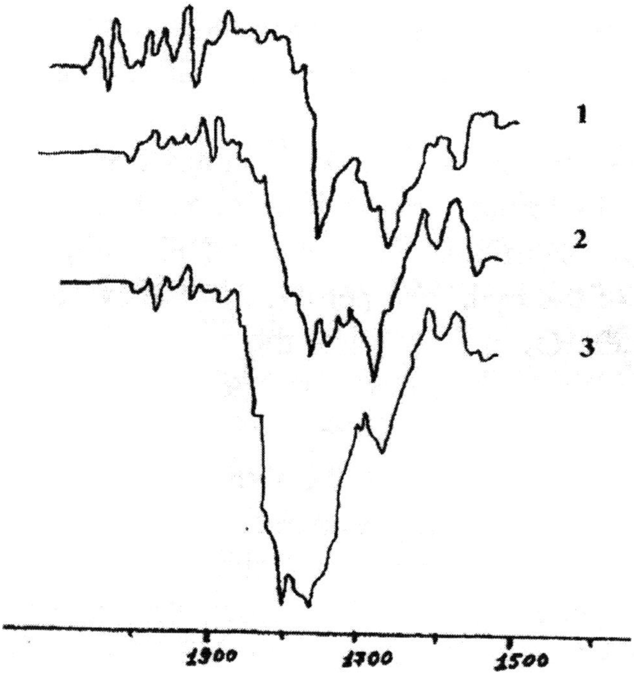

λ, cm^{-1}

Figure 2. IR-spectra of arabinogalactan [AG]: 1) original AG, 2) AG high molecular fraction after oxidative destruction at 90°C, 3) oligomer fraction of AG after oxidative destruction at 90°C

After AG oxidation by molecular oxygen in the presence of hydrogen peroxide, in the IR-spectra of the polymer and oligomer fractions, the absorbency bands of moderate and strong intensities appear in the range 1765-1750 cm^{-1} (Fig.2). The bands are characteristic for the valent vibrations of the carbonyl group. It follows from the IR-spectra that in the oligomer products, the content of the >C=O is higher compared with the oxidized macromolecules.

Conclusions

It can be concluded that at heating of the reaction mixture $AG+H_2O+H_2O_2+O_2$, the reaction of oxidation and destruction of arabinogalactan occur simultaneously. In the course of the polysaccharide oxidative destruction, the functional ization of the forming oligomers takes place.

References

[1] Ernest VGroman, Darning Gou, //*Carbohydrate Research*, 1997, 301, 69.
[2] A. 0. Arijkhoddzaev,//*Chemistry of Natural Compounds* (Russ.), 2000, 3, 185.
[3] Yu.S. Ovodov//*Bioorganic Chemistry* (Russ.), 1998, 24, 483.

[4] Alan P., Crott and Richard Bartsch, // *Tetrahedron*, 1983, 39, 1417.
[5] R.Kh. Mudarisova,N.P. Zimina, E.A. Glukhov, LA. Baschenko, V.A. Davydova, F.A. Zarudiy, Yu.B. Monakov //*Chemical Pharmaceutical Journal* (Russ.), 1997, 2, 25.
[6] G.G. Galiaskarova, I.R. Mullagaliev, Yu.B. Monakov //*Bashkir Chemical Journal* (Russ.), 1996, 3, 3.
[7] N. N. Kabalnova, V. V Shereshovets, K. Yu. Murinov, G. G. Galiaskarova, R. MuUagaliev, N. N.Krasnogorskaya, Yu. B. Monakov, //*Intern.Polymeric. Mater.* 2001, 50, 47.
[8] N. N. Kabalnova, K. Yu. Murinov, I. R. Mullagaliev, N. N. Krasnogorskaya, V V Shereshovets, Yu. B. Monakov, G. E. Zaikov, //*Journal of Applied Polymer Science*, 2001, 81, 8 75.
[9] I.R. Mullagaliev, G.G. Galiaskarova, Yu.B. *Monakov//Dokl. RAN* (Russ.), 1995, 345, 199.

INDEX

G

H

I

K

L

M

N

O

P

R